The Design Productivity Debate

Springer
London
Berlin
Heidelberg
New York
Barcelona
Budapest
Hong Kong
Milan
Paris
Santa Clara
Singapore
Tokyo

Alex H.B. Duffy (Ed.)

The Design Productivity Debate

With 83 Figures

Springer

Alex H.B. Duffy, BSc, PhD, C.Eng, C.IS.Eng, MBCS, FIED
Department of Design, Manufacture and Engineering Management
University of Strathclyde, 75 Montrose Street, Glasgow G1 1XJ, UK

ISBN-13: 978-1-4471-1540-3 e-ISBN-13: 978-1-4471-1538-0
DOI: 10.1007/978-1-4471-1538-0

British Library Cataloguing in Publication Data
The design productivity debate
1.Engineering design
I.Duffy, Alexander Hynd Black
620'.0042

Library of Congress Cataloging-in-Publication Data
The design productivity debate / Alex H.B. Duffy, ed.
p. cm.
Includes papersfrom the First International Engineering Design Debate, held Sept. 1996 in Glasgow.
Includes bibliographical references (p.).

1. Design, Industrial- -Management. 2. Engineering design- -Data processing. 3. Computer-aided design. 4. Concurrent engineering. I. Duffy, Alex H. B. (Alex Hynd Black), 1957- . II. International Engineering Design Debate (1st : 1996 : Glasgow, Scotland)
TS171.D466 1997 97-26407
658.5'75- -dc21 CIP

Softcover reprint of the hardcover 1st edition 1998

Typesetting: Camera ready by editor

69/3830-543210 Printed on acid-free paper

Contents

Part III: Concurrent Engineering

Part IV: Design Knowledge and Information

Preface

Over the past decade, with greater emphasis being placed upon shorter lead times, better quality products, reduced product costs, and greater customer satisfaction, the topic of Engineering Design has received increased interest from the industrial and academic communities. Considerable effort in the past has taken a broad view of design and has been directed at developing methodologies of the design process or alternatively building computer tools that focus upon relatively narrow aspects of design. However we seem to be little closer to finding some of the key answers to issues raised within the Engineering Design research arena and design practice.

This book contains papers and a summary report of the First International Engineering Design Debate (EDD) held in September 1996 in Glasgow, United Kingdom. The debate was directed at discussing key issues concerning the improvement of Design Productivity with a view to deriving a common understanding of the basic factors, problems and potential solutions involved. The papers reflect the work and understanding in this topic area and have been grouped under the following headings:

Design Studies

Design Development

Concurrent Engineering

Design Knowledge & Information

All papers were reviewed by two referees drawn from an international panel and one from the list of authors. They all deserve special thanks for their time, effort, pertinent comments and recommendations.

Alex H.B. Duffy
University of Strathclyde

Design Productivity

A H B Duffy
CAD Centre, University of Strathclyde, 75 Montrose Street,
Glasgow G1 1XJ, United Kingdom.

1. Introduction

Over the past decade, companies have been emphasising shorter lead times, better quality products, reduced product costs, and greater customer satisfaction. Hence, Engineering Design has received increased interest from the industrial and academic communities. These communities have focused on developing methodologies of the design process or building computer tools that focus upon relatively narrow aspects of design. However despite these efforts, we seem to be little closer to answering some of the key questions that have arisen from both engineering design practice and engineering design research.

This paper presents the deliberations of the First International Engineering Design Debate held in September 1996, Glasgow, United Kingdom. The debate was directed at discussing key issues concerning the Improvement of Design Productivity with a view of deriving a common understanding of the basic factors, problems and potential solutions involved. Thus, key questions that were addressed were:

- What is design productivity?
- How can design productivity be measured?
- What are the effective elements and how effective are they?
- How do the elements relate?

The outcomes from each of these questions are discussed in turn.

2. What is design productivity ?

After considerable debate about the nature of design productivity, the participants agreed that *efficiency* and *effectiveness* are its two key features. Efficiency was considered to be the ratio between benefits and cost:

$$Efficiency = Benefits/Costs$$

where

$$Benefits = Value\ of\ design \begin{cases} technical\ (product) \\ lead\text{-}time\ (process) \end{cases}$$

and

$$Costs = Value\ of\ capital,\ labour\ and\ overheads$$

Effectiveness, although it cannot be expressed in a simple formula, is a measure of the achievement of the desired effect or outcome.

The participants in the debate elaborated on the concepts of efficiency and effectiveness independently of the definitions above, as depicted in Figure 1. This figure shows a number of relevant elements of these two key features along with elements relevant to design productivity in general.

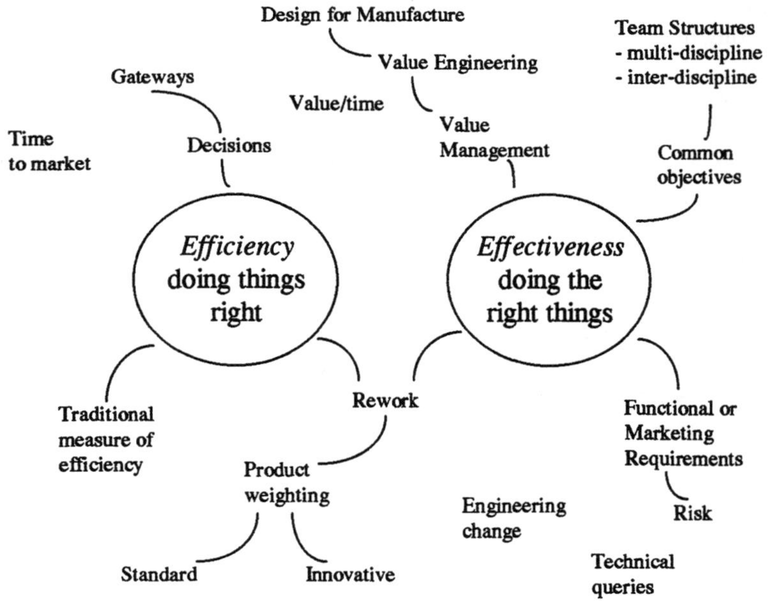

Figure 1: Elements of Efficiency and Effectiveness

But can design productivity be an all encompassing concept or is it concerned with an individual, a group, a project, or a company? The result was a consideration of different "levels" of productivity, as shown below. A hypothesis, that *Efficiency* is more apparent and measurable at the lower level and *Effectiveness* more so at the higher level, was presented.

a) Business Productivity

b) Project Productivity

c) Team Productivity - Sub-systems

d) Individual Productivity

e) Product or Output measures (e.g. hours/task)

For a particular project, group, or individual the following factors were considered to have an affect on the degree of productivity:

Project	Team	Individual
Customer needs Information management Plan Resource allocation	Communication Formation/structure Leadership Role allocation Trust	Awareness Clear tasks and goals Experience Innovation/ creativity/ ideas generation Knowledge Motivation

As a result of the above considerations, design productivity was defined as:

The efficiency of production of a design solution, within a business context, that is effective to the overall requirements

where *efficiency* relates to the process and *effectiveness* the product's development.

3. How can design productivity be measured ?

This section presents the results of a general discussion on the measurement of design productivity and potential barriers for enhancing productivity. The measures are generic in nature and so can be applied to different elements of productivity depending on the element and the desired effect. Some participants argued that while measures can be useful, caution should be taken as some measures can be counterproductive. The focus of this part of the debate was on defining the key elements for measuring and improving design productivity.

Some generic units of measure were listed as:

- Appropriateness/meet the demands/needs
- Cost
- Labour: man hours
- Number of different aspects
- Technical considerations:
  - fulfilment of actual need
  - fulfilment of technical requirements
  - fulfilment of needs of subsequent life-cycle phases
  - reuse of knowledge
  - reuse of product/design robustness
  - value if sub-contracted
- Time

The table below shows particular dimensions of productivity that can be measured for each design phase. For example, the main elements of productivity apply throughout, but particular elements may be more apparent during certain phases. That is, idea generation may be more apparent at the concept rather than at the detailed design phase. Further, these dimensions of productivity may be measured by say the number of sales or degree of fulfilling the brief.

Phase	Productivity	Measures (Efficiency/Effectiveness)
Concept	Ideas	Sales Fulfilment of Brief (Plans/Spec.)
Schema	Resources	Lead Time Cost
Detail	Added Value	Rework Reliability (product & team)
	Risk Reduction	Risk Assessment Customer claims

During the discussions it became apparent that there are numerous barriers to enhancing productivity. The following list highlights some of the ones proposed during the discussions:

- Approach: converge on a design solution too soon, poor selection and use of tools, late consideration of constraints, inadequate documentation (design histories), lack of a process model.
- Communication: distance, infrastructure, organisation, different language styles.
- Experiences: constrained by prior experiences, too much (mind set).
- Expertise: mismatch
- Individuals: decision making ability, skills, problem solving ability, poor motivation, poor learning ability.
- Inflexibility: reluctance to change.
- Information: lack of, incorrect, too much, changing.
- Knowledge: lack of knowledge, poor training, inappropriate education, lack of documentation.
- Management: process too rigid, poor decision making, lack of focus, poor leadership, bad planning, uncoordinated, untimely decisions.
- Planning: poor, lack of.
- Product: designing the wrong solution, unclear specification, too many conflicting goals, poorly understood goals.
- Resources: lack of, insufficient funding, wrong or poor tools, lack or unsuitable people, inappropriate allocation.
- Teamwork: poor cohesion, lack of focus and direction, “in-fighting”, clash of personalities.
- Technical: poor equipment, tools not available.

Some participants argued that these barriers, when inverted or aligned to enhance productivity, are themselves elements of productivity.

4. What are the effective elements and how effective are they?

At a general level productivity may be considered to reflect the outcome of a company’s business. Examples of productivity at this level are the number of products, profit margin, market share, competitive standing, and the development time, cost and quality of the product.

Within the product development process itself, the participants in the debate generated and grouped a number of elements that were considered to be effective in design productivity, as shown in the two tables below. The question of "How effective are they" was not addressed.

CONTEXT	ORGANISATION STRUCTURE	DESIGN/ DEVELOPMENT PROCESS	PRODUCT
• Competitors • Internationalism • Market/ Customers/Users • Politics • Suppliers	• Allocation of resources • Flexibility/ Degree of functional integration Location • Reporting procedure • Teamwork – creative environment – co-location – communication/ technology	• Availability of resources (people, skills, technology) • Degree of emphasis • Goal orientation • Methodology; DFX's (supply chain integration) • Planning/Review of design activities • Stages • Task orientation • Technology and tools	• Complexity • Life-cycle issues (easy-to-X) • Maturity • Production volume • Robustness • Scientific content

GOALS	TASKS	COMPANY /ORGANISATION	
• Determine together with elements	• Determine together with elements	People: • Commitment & Motivation • Communication ability • Competitiveness • Degree of being empowered • Heuristic competence; problem solving knowledge • Individual style, leadership quality, "people skills" etc. • Knowledge, Expertise and Experience • Rewards; remuneration, motivation, social issues Role • Task allocation • Training & Skills	Equipment: • Accessibility, ease of use • Appropriateness of tools, technologies, etc. • Availability • Computing/ Communication • Integration • Knowledge • Maintenance • Quality/ state-of-the-art

Table 1: Elements of design productivity

There are obviously different "trade offs" between the above elements depending upon the market place, type of industry, product, organisation, etc.

5. How do the elements relate?

To attempt to uncover the relationships among the elements, the debate participants had to determine the nature of design and design productivity within the business context.

The drive to enhance design productivity can influence the goals of product development which themselves influence the goals and achievements of the business:

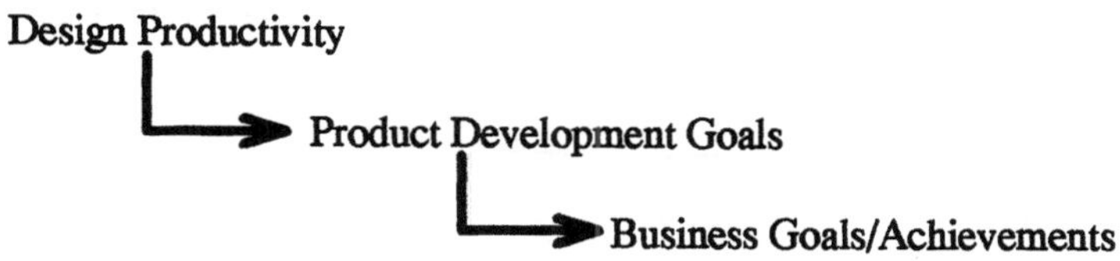

Alternatively the business goals can be considered to determine the set of possible elements required to be investigated to enhance design productivity.

The participants in the debate agreed that design is an encompassing activity with particular needs as input, with a product description as output, and with goals, constraints, and resources as influential factors (Figure 2).

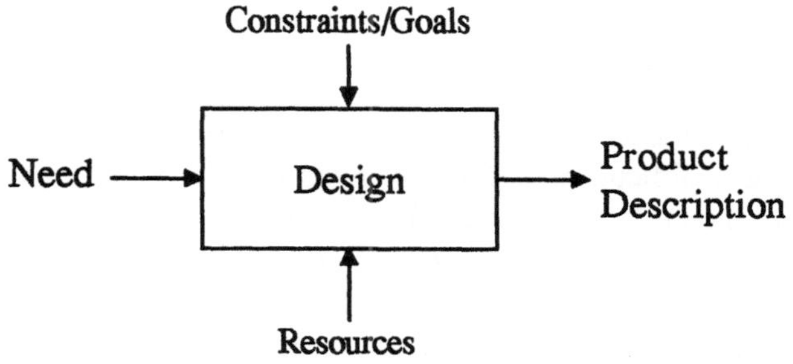

Figure 2: Design activity

Figure 3 presents the relationships among the elements and their effect on enhancing design productivity. In this figure, each design activity influences others to some degree. To enhance design productivity, each productivity measure (listed in Section 3) can be assigned to sub-goals with corresponding tasks to be addressed for each activity. Metrics could be pre-defined for elements presented in Section 4 and used to measure the performance of the activities with respect to the identified goals and tasks. The application of these metrics should reflect an improvement in efficiency (%) and/or effectiveness (σ) thus ensuring an overall performance improvement within and across design activities.

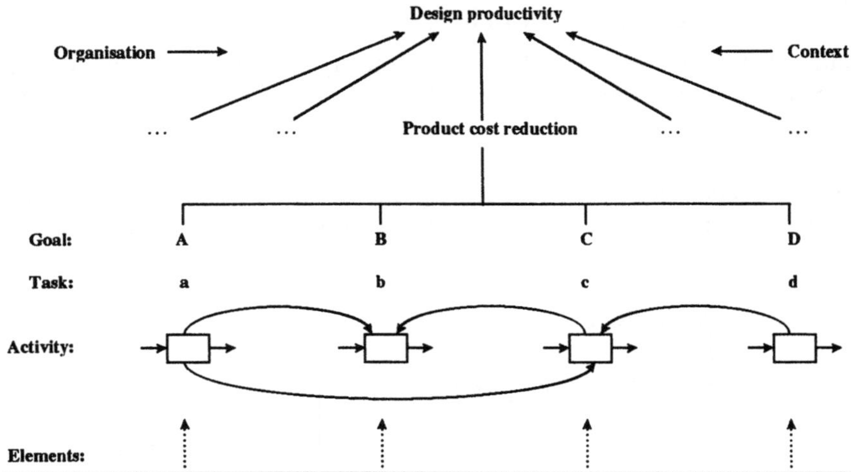

Process —	%/σ	%/σ	%/σ	%/σ
Product —	%/σ	%/σ	%/σ	%/σ
People —	%/σ	%/σ	%/σ	%/σ
Equipment —	%/σ	%/σ	%/σ	%/σ

Figure 3: How the elements relate to enhancing design productivity

6. Conclusion

This paper presents the consensus of the participants in the design debate concerning a difficult and still ill-defined topic. While many of our research efforts are directed at improving design productivity, it is remarkable how little we know of this topic. The debate was directed at gaining not only a recognition of this important issue but also an insight into the key features of design productivity. While this was achieved to some degree, there are many questions that need to be answered before we can fully define the true nature of design productivity.

Having said this, the main findings from the debate can be considered to be:

- Design Productivity was considered to consist of two key elements - Efficiency and Effectiveness.

- Design Productivity was defined as *the efficiency of production of a design solution, within a business context, that is effective to the overall requirements.*

- A number of general dimensions for measuring, along with barriers to enhance, productivity were listed.

- Key elements were listed and an overall structure of how these elements relate to enhancing design productivity within a business and design activity was presented.

7. Acknowledgements

This document reflects the cumulative effort of all the participants of EDD'96 and in no way should be attributed to or may be considered to express the opinion or views of the author or indeed any single participant. The author is grateful for comments and suggestions from Professor S Finger and Mr F O'Donnell on a draft version of this paper.

The participants who contributed to the findings of the debate are:

Prof. M M Andreasen, Technical University of Denmark, DK.
Dr S Austin, Loughborough University of Technology, UK.
Dr P Bedke-Schaub, Universitat Bamberg, D.
Dr. L Blessing, University of Cambridge, UK.
Dr M Cantamessa, Politecnico di Tornio, I.
Mr C Cointe, INRIA, F.
Dr A Court, Imperial College of Science, Technology & Medicine, UK.
Dr A Duffy, University of Strathclyde, UK.
Prof. S Finger, Carnegie Mellon University, USA.
Dipl-Ing. E Frankenberger, Technical University of Darmstadt, D.
Dr F Lettice, Cranfield University, UK.
Prof. K J MacCallum, University of Strathclyde, UK.
Dr E McMahon, University of Tennessee at Chattanooga, USA.
Dr S Minneman, Xerox PARC, Palo Alto CA, USA.
Dr M Murray, AMEC Design and Management, UK.
Prof. M Norell, Kungl Tekniska Hogkolan, S.
Mr R Parsons, University of Plymouth, UK.
Prof. N Roozenburg, Delft Universityof Technology, The Netherlands.
Prof. T Tomiyama, University of Tokyo, J.

Part I
Design Studies

Investigating Productivity in Engineering Design: a Theoretical and Empirical Perspective

Marco Cantamessa
Dipartimento di Sistemi di Produzione ed Economia dell'Azienda
Politecnico di Torino, Corso Duca degli Abruzzi 24 - I 10129 Torino (Italy)

Abstract

Engineering design activities in industrial companies are very complex, and complex also are the relationships with the firm's economic, technical and social environment. It is therefore quite difficult to apply the concept of productivity to engineering design, especially if the objective is to improve it with the many techniques which have been proposed during recent years. The paper, based upon theoretical reflections and empirical evidence, proposes a broad concept of design productivity which takes the company's specific product development objectives into account, as well as the degree with which the design function contributes to satisfying them. It is also shown that techniques for product development induce significant tradeoffs and synergies between such objectives; this suggests that a prudent stance be taken when evaluating the efficacy of specific tools, their value seeming to lie more in the subjective *mode d'emploi* with which companies incorporate them into operations, rather than in their objective worth.

1. Introduction

The attention which both academia and industry have paid to engineering design and product development during recent years doesn't probably need to be emphasized in this paper. Among the expressions of such attention, it is possible to mention the well-known empirical studies performed at MIT and Harvard University at the turn of the decade [1,2]: such investigations, besides bringing into focus the importance of manufacturing operations for industrial competitiveness, have specifically emphasized product development's central role. As a result of this recent focus upon engineering design, basic and applied research work on the subject has recently become abundant, and industry has started to look with interest at the generous proposal of new tools for supporting product development: both "hard" technology (principally Computer-aided systems such as CAD, CAM, CAE, etc.), "soft" methodologies (recent ones such as QFD and DFMA, together with older ones such as Value Analysis) and, principles)[1]. Indeed, it seems that many expectations which industry had cast upon Advanced Manufacturing Technology for gaining competitiveness during the '80s, have recently been reversed upon the topic of product development.

Due to the recent excitement surrounding engineering design, it seems to be quite meaningful to research upon industrial experience in the innovation of product development practices, so that academic research efforts and industry's attitude in the management of innovation may be appropriately directed. Such an issue appears to be quite important, since much research work is still in progress, and a quick glance at projects being carried out at various institutions worldwide shows that a new generation of innovative techniques will be made available to industry in the near future (a comprehensive survey may be found in [3]).

In the context of studying the adoption and utilization of techniques for product development, the topic of *design productivity* which is being addressed in this Debate is a central one indeed, since most of these techniques are actually intended to improve productivity. However, tackling this issue unveils a complexity which may not be neglected, if not by settling for a simplistic picture of engineering design in industry. For example, it would be quite questionable to interpret the term "productivity" as a simple ratio between the sheer volume of design outputs (e.g. number of projects performed) and the resources spent for producing them (typically economic ones, such as capital and labor). Due to the peculiarity of design (which is an information and knowledge-intensive activity) and its special role in the manufacturing company (especially small and medium-sized ones), not only the volume of design outputs should be considered but also their quality, where by quality one may intend the degree with which such outputs contribute to satisfying the competitive strategy of the firm.

So, the problem is to identify which should be considered as outputs and inputs of the design activity in an industrial company, and to define their quantity and quality. This may not be considered an easy task, because of the apparently inextricable relationships between the technological, methodological, managerial, economic and social factors which are related to engineering activities. Sidestepping such a complexity would be a mistake, because most recent research efforts are actually concerned with developing tools which take these factors into account altogether: one can become aware of this simply by looking at the list of issues which are covered in this book.

The paper aims, starting from both a theoretical and an empirical perspective, to bring into focus the subject of design productivity in manufacturing companies and to discuss the impact which support techniques may have upon it. In the following section, the topic will be introduced by presenting and discussing at some length an abstract model of design and product development in manufacturing companies. The subsequent section will then propose a set of *paradigms* of engineering design, which may be used in order to define the different technical objectives which companies choose for planning their product

---

1 In the following, the term "techniques" will be used to collectively indicate all these tools for supporting product development, regardless of their different nature.

development practices. Finally, in the last section, quantitative analysis of a data set obtained through a recent survey will give empirical support to the discussion. As a remark, readers may have noticed that, in the previous sentences, the terms "engineering design" and "product development" have been used in a nearly interchangeable way; it is not inaccuracy behind this choice, but rather the consciousness that, in the small and medium size enterprises upon which this research is focused, product development almost entirely coincides with engineering design activities.

2. A Theoretical Model of the Design Function of the Firm

The objective of the design function in a manufacturing company is evidently that of transforming initial marketing or technical specifications into the full set of information which defines the product over its life-cycle, from manufacturing to its final disposal. The design function's operation will therefore be thought of as a form of information processing. In order to model the design function of the firm, it is convenient to differentiate between *flows* (in our case, information and labor) and *stocks* of assets, which allow the production of design. The peculiarity of stocks is that they aren't consumed by production, but are subject to phenomena such as accumulation and deterioration. Stocks may be both capital assets (e.g. CAD workstations, lab equipment, etc.) and knowledge assets, which may conveniently be categorized into technical-scientific and methodological knowledge. Concerning the former, it is evident that engineering design requires a wealth of technical-scientific knowledge, which is connected to the various disciplines of engineering and depends upon the specific technologies which are incorporated in the company's products (e.g. mechanical, electrical, software, etc.). Methodological knowledge, instead, is the domain-independent know-how of what design and product development *are*, and of how they should be performed and managed. Using Vincente's [4] classification of the different forms of engineering design knowledge, it is equivalent to what the author terms "design instrumentalities". Knowledge assets are subject to growth both from the outside (e.g. attending to conferences, reading books, etc.), as well as an outcome of the design activities which have been performed, through a learning mechanism.

The obvious way with which the design function of the firm may be acted upon is by changing it explicitly, for example by adopting support techniques which will affect its components; this may be termed *meta*design, since it deals with "designing the design function of the firm". However, the same objective may be conveniently pursued by exploiting the above mentioned learning mechanism.

The components of the model in Figure 1 may be qualitatively characterized by introducing attributes for them; the model may then be used for reasoning, from a theoretical point of view, about the consequences of adopting design support techniques. In fact, when a company adopts a technique, this will change

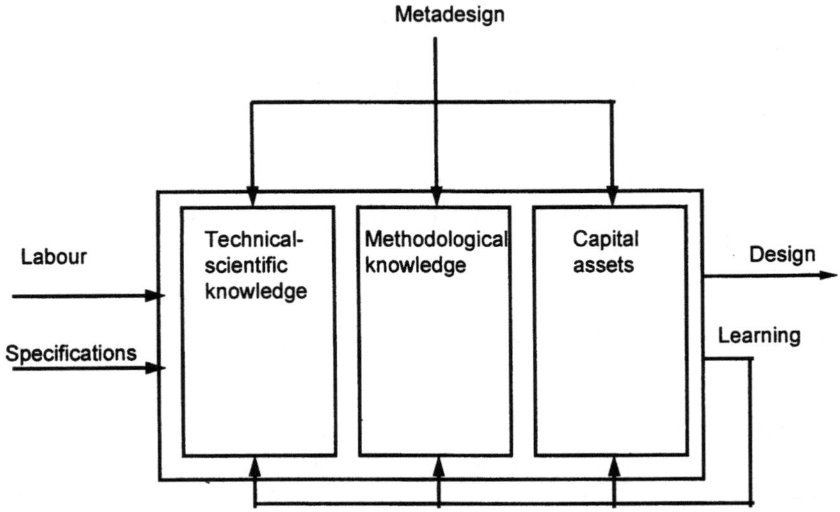

Fig. 1. *An abstract model of the design function of the firm.*

its design assets and, therefore, modify the design function and the model which represents it. It is outside the scope of this paper to systematically discuss components' attributes, but a simple example may probably help to clarify some relevant concepts (when some of these attributes will eventually be introduced, they will be written in italics):

> "The ACME company wishes to overcome a rather specific technical problem in the design of widgets; one idea which comes into the mind of ACME's technical director is to hire a specialized engineer able to solve the problem thanks to his ability with, say, modeling problems with partial difference equations, solving them, and then using the solution for optimizing design variables. Following this idea, the technical-scientific knowledge of ACME's design function wouldn't only increase in *magnitude*, but also in *specialization*. However, another solution could be to acquire a complex and costly CAE system, and using it in order to simulate a host of different solutions until a satisfactory one would come out. In this latter case we would see the capital assets of the company increase in *magnitude* and, if such a CAE system is unable to read and write product models with a standard format, their degree of *integration* would diminish. Concerning technical-scientific knowledge, one could imagine it to remain unchanged both in *magnitude* and in *specialization*, since existing engineers could probably be easily trained to operate the system. Now, let us suppose that ACME also has the objective of developing innovative products: this second requirement wouldn't probably benefit from a

> excessive degree of technical-scientific *specialization*, and this factor would therefore make the latter solution the best one".

Of course, readers may easily draw from their own experience and elaborate many other examples involving other components of the model and their attributes. This simple example anyway shows that, depending upon the company's specific objectives, a choice will be made among the different techniques available; these techniques will then affect the design function of the firm with a set of intended and unintended side effects. Collectively, the adoption of techniques leads to an extremely complex pattern of synergies and tradeoffs between objectives, which have to be considered when discussing the productivity of design and the choice of techniques for increasing it. It should be stressed that, in principle, there is no direct synergy or tradeoff between objectives; on the contrary, synergies and tradeoffs emerge because of the nature of the specific techniques which are adopted for satisfying them. Of course, the most desirable techniques for supporting product development therefore are those which maximize the synergies while minimizing the tradeoffs.

The existence of synergies and tradeoffs has important consequences concerning design productivity: first, the effects produced by the elements which make up the design function are not additive, but are the outcome of quite complex interactions. So, the entire picture of a company's design function has always to be taken up as a whole whenever one wishes to evaluate the actual productivity and find ways for improving it. As a second consequence, it is quite hard to hypothesize that an absolutely optimal configuration for the design function of a firm may exist; on the contrary, specific objectives and available techniques must be carefully matched in order to find the configuration which will give, at a certain moment in time, the best results.

3. Characterizing Design Objectives and Support Techniques

In the previous section it has been shown that, in studying design activities, it is important to jointly consider design objectives and support techniques, altogether with their mutual relationships. Now, in order to proceed further, it becomes necessary to identify relevant objectives and, even more, to categorize the many existing techniques; therefore, an appropriate characterization for these two elements has to be found. To this purpose, a powerful concept which may be borrowed from the field of evolutionary economics is that of "paradigm" [5]. A paradigm may be broadly defined as a distinct collection of technical problems and related solutions; paradigms are characterized by their historical derivation, and thus advance over time by following "evolutive trajectories". Therefore, the identification of paradigms of engineering design calls for an investigation upon the historical emergence and evolution of distinct clusters of problems and technical solutions in engineering and product development.

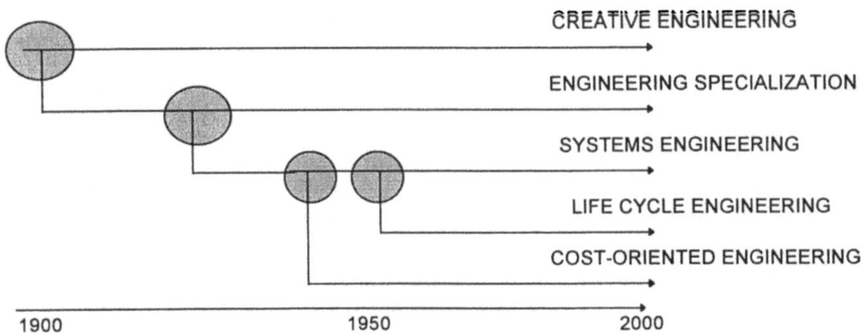

Fig. 2. *Evolutive trajectories for the paradigms of engineering design and product development.*

The paradigms which will be proposed are shown in Figure 2. It has to be mentioned that the analysis which have led to their identification hasn't yet been performed with the rigorous criteria and procedures a historian of technology would probably ask for. Rather, it has been based upon the basic knowledge upon design support techniques which is obtained by using textbooks, academic and trade journals as sources. Nonetheless, such a framework seems to be quite powerful: other non-chronological categorizations had also been considered previously during the research, but did not show a sufficient degree of objectiveness.

It is important to remark that Figure 2 indicates the emergence of different engineering paradigms, defined by a set of distinct technical problems and "toolboxes" of available support techniques, but not their adoption by companies. It is therefore possible to see Figure 2 as a time-based representation of the possible choices (or "best practices") which have been offered to companies over time. At any moment, a specific company may therefore decide to pursue the objectives related to one or more paradigms, and do so by adopting some of the available techniques which pertain to them. Incidentally, the view of paradigms as collections of "best practices", could allow to interpret them (at least in part) as coincident with the professional figures which emerge after being educated according to different training curricula.

The chronological derivation of the five proposed paradigms and the discussion of the techniques related to each of them cannot be exposed at length because of the space it would require, nor would it be in the scope of the paper. So, only a very short summary will be introduced in the following, just to acquaint the reader with the framework and lead him to the empirical analysis proposed in the next section of the paper.

Creative engineering represents the objective of designing in order to generate radical product innovations, breaking free from pre-existing solutions. Related in some way to technical invention, this may be seen as the original form of engineering design: incidentally, most design theories give a great importance to conceptual design, which is central to this paradigm. Conversely, routine design is known to be much more common in industrial practice. It is possible to associate to this paradigm the tools coming from theories of engineering design (e.g. systematic design techniques), as well as methods for supporting creativity (e.g. brainstorming and synectics).

Engineering specialization is related to the objective of improving the scientific content of products and optimizing specific aspects of their behavior (e.g. the aerodynamic drag of a car body or the fatigue resistance of a crankshaft). This objective is pursued by adopting techniques and tools which are specific to different engineering domains, and generally depend upon quite sophisticated analytical or numeric solution methods. From a historical point of view, one can observe the emergence of this paradigm at the turn of the century, when "new" products (such as electrical machinery or internal combustion engines) forced to go beyond primeval engineers' reliance on intuition and "rules of thumb" analytical thinking.

Systems engineering represents the objective of designing in order to develop complex products consisting of disparate modules combining different technologies. The emergence of complex products such as those in the aerospace industry, which were characterized by heterogeneous technologies - and the simultaneous "deepening" of the engineering knowledge required by each technology - may be seen as the events which originated the paradigm of systems engineering. The realization of such products requires large amounts of work in defining and negotiating specifications at the different levels of the product hierarchy and coordinating the resources which participate in the design process. Among the techniques which literature commonly attributes to the systems engineer's toolbox it is possible to mention, from the technical point of view functional analysis, value analysis and engineering, QFD; from the organizational point of view, it is possible to mention project management and design review techniques. Due to the complexity of modern artifacts, the systems engineering paradigm permeates most of the design activities which can be found in industry; moreover, it is strongly connected with organizational issues within the company (the structure of design departments generally follows the architecture of complex products) and the organization of interfirm relationships.

Life cycle engineering has the objective of designing products in order to consider not only their technical functionality, but also all of the aspects which pertain to their life-cycle, from manufacturing through their removal from service. Although life-cycle engineering emerged in the aerospace industry together with systems engineering, there are important differences between the two. First, technical skills are different since, while systems engineers are generalists by definition, life-cycle engineering often leads to distinct specializations and "toolboxes" of techniques (e.g. analysts trained in reliability, ergonomics, or environmental impact on the side of skills, while FMEA, FTA and DFX may be mentioned on the side of methods). Moreover, while systems engineering generally leads to structured and formal information flow and to hierarchical organizational structures, the variety of issues related to life-cycle engineering favors quick and informal information flow and organizational forms crossing functional boundaries (e.g. concurrent engineering principles, interfunctional project teams).

Cost-oriented engineering has the objective of supporting design in order to achieve the optimal static efficiency, i.e. to perform repetitive design activities in the most economically efficient way (it should be noted that the aim is to reduce the cost of the design project and not that of the product). It is within this paradigm that the concept of simple productivity finds its source. From a technical point of view, the tools used for cost-oriented engineering are the automation of engineering activities and the re-use of existing data and designs, generally with an intensive use of Information Technology.

An alternative view of the proposed set of engineering design paradigms is given in Figure 3, which shows the evolution of support techniques within the framework defined by the paradigms. Specifically, the dotted arrows in the picture indicate the techniques which were originally developed in order to serve the needs of a specific paradigm and have then evolved into newer ones in order to be

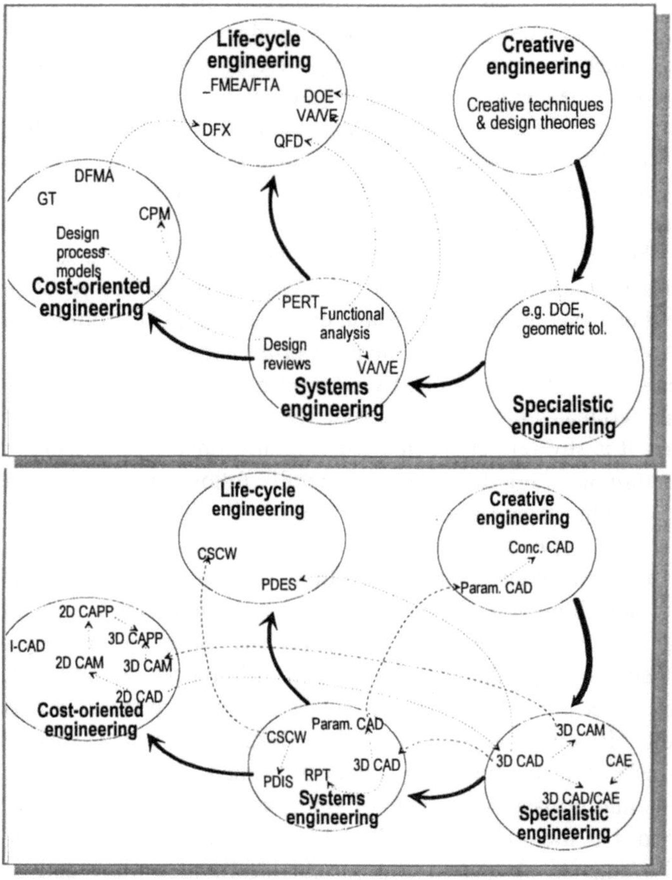

Fig. 3. *The evolution of design support methodologies (top) and technologies (bottom).*

used within another paradigm. Dashed arrows, instead, indicate techniques for which the utilization within different paradigms has not required any transformation, but simply a change in their usage. Incidentally, it may be observed how such latter movements are relatively rare for methodologies, but are far more frequent for computer-aided technologies. This consideration may help to support the hypothesis that technological determinism does not apply to these technologies.

4. Effects of Design Support Techniques. An Empirical Analysis

The previous sections of this paper have introduced the concept that the effects of support techniques upon design productivity have to be studied from a rather broad perspective: the entirety of firm's product development function should be taken into account, including its knowledge and capital assets and its specific objectives. Such a thesis was based upon a theoretical discussion; the objective of this last section of the paper is now to support it with some empirical evidence.

After briefly presenting the structure of the underlying research project, the following subsections will introduce three specific arguments. The two first ones will deal with cases in which significant interrelationships among techniques have been found, the former showing the existence of a synergical effect, and the latter of a tradeoff. Finally, the discussion will concentrate upon the central role which organization has, not only in performing design activities, but also in the adoption and implementation of support techniques. It will then be suggested that organizational factors could be among the factors which contribute to explain the complex relationships existing among objectives, techniques and effects.

4.1 Methodology

The empirical research upon which this section is based derives from the work being carried out by a multidisciplinary research group on innovation in product development. The group was established in 1994 at the author's institution under the sponsorship of the local Chamber of Commerce, and has the objective of performing empirical investigations on engineering design and product development jointly from engineering, organizational and economic points of view. Group membership is therefore quite composite and includes researchers from these different areas.

After a preliminary phase dealing with bibliographical research [6], which has provided the basis for the previous theoretical discussion, a second step consisting of field activity was performed. During this stage, the group performed more than twenty day-long visits and interviews to manufacturing companies, drafted, validated and discussed the related case studies; this allowed to put into focus the main issues faced by companies in their experience in innovating product development practices [7]. The research finally entered its final phase, based upon a mail survey [8]: a rather detailed and broad questionnaire was designed and sent to the universe of companies considered, consisting of the 620 discrete parts manufacturing firms located in the region, with more than 50 employees and less than 5000. The questionnaire was a 32 page booklet containing about 200 questions covering various aspects related to product development; in order to ensure a satisfactory response rate and quality, help was provided to companies both through telephone assistance and on-site visits. The response rate has been quite satisfactory and has led to a final sample of 98 companies, which has been found to be representative of the universe under consideration concerning product classification, and slightly biased towards larger companies concerning firms' size. A strong orientation towards the automotive industry also characterizes the region being studied, because of the local presence of the largest Italian car manufacturer and of many of its suppliers.

In order to proceed further in the discussion, it is now necessary to develop, starting from the previous theoretical discussion, a model which may support the quantitative analysis of the adoption of design support techniques and of their effects. Specifically, the objective is to perform an cross-sectional kind of analysis. The proposed model is shown in Figure 4, and is based upon four macro-items, namely *context*, *objectives*, *techniques* and *effects*, together with their reciprocal influences. Context includes variables which may to some extent be considered as exogenous, and describe the company from an "objective" point of view (e.g. its size, the nature of its products, its possible belonging to large multinational firms, the kind of relationship entertained with customers, etc.). Objectives represent which of the previously introduced design paradigms are being pursued by the company. These may thought to be partly induced by context and partly dependent upon an explicit choice made by the company. Techniques represent the set of technologies and methodologies the company has adopted, based upon its objectives and influenced both by context and by the prior adoption of other techniques. Finally, effects represent the eventual increase in operational performance and/or competitive position.

Concerning data analysis, the raw data collected during the survey is mostly of categorical nature, and has been reduced to a binary format for homogeneity; analyzing a data set of this kind therefore requires the usage of appropriate

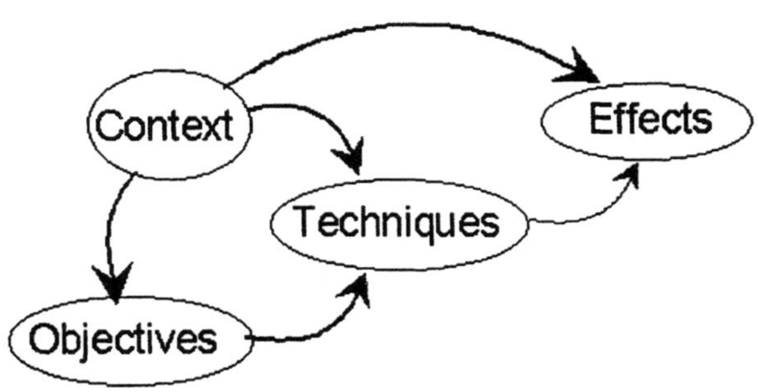

Fig. 4. *A model for interpreting the adoption of support techniques.*

techniques. As a preliminary step, contingency tables between couples of variables belonging to the four categories of items described above have been built, and significant associations spotted by using both the Lambda and the Goodman-Kruskal Gamma statistics. However, such a method doesn't allow to interpret the phenomena satisfactorily, since it doesn't provide a global view upon the pattern of associations between variables, but only upon separate couples of them. In other words, most variables do appear to be significantly associated among themselves but, in many cases, the true reason behind the association existing between variables A and B is that both of them happen to be associated with a third variable C

Loglinear analysis has therefore been used in order to find significant effects in accordance to the model shown in Figure 4. The loglinear models have been initially built by including those variables for which a statistically significant degree of association had been found with contingency tables. Models have then been modified by eliminating variables whenever possible and reintroducing them when required; the criterion used for variable selection has been the achievement of acceptable values in the models' goodness-of-it measures, expressed by the Likelihood Ratio Chi square and the Pearson Chi square. For the sake of simplicity, loglinear models have been limited to main, second-order and, only when needed, third-order effects. In order to limit the length of the coming discussion, the meaning of variables used in loglinear models will be briefly described, but their derivation from the raw data set will be given for granted (details are given in [8]).

4.2 Techniques are not additive: the synergy between 3D CAD and project management methods

Among design support technologies, some interesting patterns may be obtained by analyzing the adoption and usage of 3D CAD in the companies present in the sample. CAD based upon solid modellers is a topic which has been studied quite widely in empirical research [9-11] and lends itself quite well to looking at the multi-faceted role of design support techniques. At the date of December '94, the status of CAD adopters in the sample is shown in Table 1. Loglinear analysis of the adoption of CAD 3D involving context, objectives and effects yields the results in Table 2. The table reports the list of variables included in the model, the log odds ratio related to the two- and three-way effects whose value appeared to be significantly different from zero at 95% confidence level and, finally, the model's goodness of fit[2].

---

[2] as a reminder concerning the statistics being used in the tables, the odds O(E) of an event E is defined as $O(E) = P(E)/(1 - P(E))$, where P(E) is the probability of E. In the case of binary variables A and B, the log odds ratio L(A,B) is a measure of association which is given by the logarithm of the ratio between the odds of A = 1, given B = 1, and of A = 1, given B = 0. In other words, this means the odds of A = 1 are exp(L(A,B)) greater when B = 1 than when B = 0. Concerning goodness of fit, the likelihood ratio and Pearson Chi square measures represent the α-level at which one can reject the null hypothesis that the

Table 1. *Status of CAD adopters in the sample.*

non respondents	no CAD	2D CAD	3D CAD
3 (3.1 %)	17 (17.3 %)	36 (36.7 %)	42 (42%)

Table 2. *Loglinear model representing the adoption of 3D CAD.*

Variables in the model	Context : large company, ownership by a multinational company Objectives: engineering specialization paradigm Techniques: 3D CAD Effects : project duration reduced by more than 20%	
Effect	**Description**	**Parameter (log odds ratio)**
Two-way effects	engineering specialization paradigm * 3D CAD 3D·CAD * project duration reduced by more than 20%	0.27 0.35
Three-way effects	none	-
Likelihood Ratio Chi Square = 26.82256 DF = 49 P = .996		
Pearson Chi Square = 28.04820 DF = 49 P = .993		

The table shows that context variables in the model aren't significantly associated with the adoption of 3D CAD; rather, the characteristic which most frequently appears in adopting companies is the objective of developing products with a high degree of engineering content. Surprisingly enough, however, it is found that the principal effect which may be observed in association to 3D CAD usage is not the design of superior products, but the shortening of development lead time. It therefore appears that this technique has an important side effect, although a desirable one. It may now be interesting to analyze the relationship, if any, of 3D CAD with other techniques which also contribute to obtaining this same unexpected result. Project management methods (i.e. PERT/CPM) are good candidates for this, and the loglinear model which involves both 3D CAD and project management tools gives the results in Table 3.

In this model, the previously obtained effects related to 3D CAD appear to be substantially confirmed. For what concerns project management techniques, context appears to cast a significant influence upon their adoption (these methods happen to be used by larger companies and in association to continuing education programs) and their effect is - indeed - the reduction of development time. However, the interesting evidence is that a significant combined (three-way) effect with 3D CAD may also be found, which shows the presence of a synergical effect between these two techniques.

loglinear model does **not** fit the data. Values exceeding 0.7 are usually considered acceptable [12].

Table 3. *Loglinear model representing the adoption of 3D CAD and project management.*

Variables in the model	Context : large company, ownership by a multinational company, continuing education programs Objectives: engineering specialization paradigm Techniques: 3D CAD, project management techniques Effects : project duration reduced by more than 20%	
Effect	**Description**	**Parameter (log odds ratio)**
Two-way effects	engineering specialization paradigm * 3D CAD	0.21
	large company * project mgmt techniques	0.52
	continuous education programs * project mgmt techniques	0.28
	project mgmt techniques * project duration reduced by more than 20%	0.46
	3D CAD * project duration reduced by more than 20%	0.30
Three-way effects	3D CAD * project mgmt techniques * project duration reduced by more than 20%	0.19
Likelihood Ratio Chi Square = 58.67968 DF = 99 P = 1.000		
Pearson Chi Square = 82.09543 DF = 99 P = .891		

A word of warning is now necessary at this point: these results (as well as the ones which will be presented in the following) are intended to be descriptive, and their aim is no more than to demonstrate the existence of complex relationships between the items shown in the conceptual model of Figure 4. These relationships could be causally interpreted only by developing a comprehensive framework for explaining the nature of the relationships existing between design support techniques and design activities. In other words, research questions which should be placed are: "which are the factors which enable the firm to adopt and implement technique X, and how do they work?", and "thanks to what mechanisms does technique Y yield effect K?".

4.3 Techniques are not additive: the tradeoff between DFX and interfunctional groups

Another interesting interaction between techniques emerges by analyzing the methods with which companies face the problem of product life-cycle issues in design. During the research, nine life-cycle factors have been examined, namely reliability, maintainability, safety, ergonomics, aesthetics, operational costs during useful life and, finally, environmental impact both during operation and at the time of disposal. Aspects connected to product realization (i.e. manufacture, assembly, etc.) have also been considered, but will not be taken into account here.

As a starting point, it is known from literature that widely recommended techniques for dealing with life-cycle issues are organizational tools such as interfunctional teams and methodologies such as DFMA. Concerning organizational solutions, their distribution is given in Table 4; during the following analysis the focus will be placed upon the rightmost category, related to formal interfunctional groups. Regarding methodologies, a number of companies in the sample have collected and codified parts of life-cycle knowledge and written design guidelines and/or checklists for directing designers' decisions (incidentally, though using a similar method, the acronym DFMA happens to be unknown to most of them). The status of adopters in the sample is shown in Table 5; results sum up to more than 100% because some companies use both design guidelines and checklists. In the following, a company will be considered to be an adopter of DFX if either of the two methods is used.

The loglinear model which studies the adoption of DFX and interfunctional groups is shown in Table 6. As expected, both are associated with companies pursuing the life-cycle paradigm. Now, the interesting fact is that a negative three-way effect exists between the two techniques and their common objective; this tradeoff could be interpreted as if companies perceived DFX and interfunctional groups to be substitutes, the former pushing towards the codification of knowledge (in a way, towards "automating" the hints human experts give within an interfunctional group), and the latter allowing a free and informal flow of such knowledge.

4.4 The role of organization

After having detected some of the complex and interrelated effects which design support techniques have upon product development activities, it is now quite natural to investigate upon which may be the relationships between this phenomenon and organizational factors. In fact, since the first steps performed during the field research which prepared the postal survey, a major principle which seemed to be a constant across all the firms visited was the strong influence organizational factors appeared to have upon product development and upon the adoption of support techniques. Both interfirm relationships and intrafirm organization appeared to have a major role in characterizing design activities, objectives and the adoption of support techniques. Quantitative analysis of the

Table 4. *Status of organizational designs' adoption in the sample.*

Non respondents	Designers work alone within functional structures	Designers work in groups within functional structures
3 (3.1 %)	22 (22.4%)	6 (6.1%)
No group, but project managers act as liaisons between functions	Informal inter-functional group (mainly small firms)	Formal inter-functional group
3 (3.1 %)	17 (17.3%)	47 (48 %)

Table 5. *Status of DFX adopters in the sample.*

no form of DFX	design guidelines	checklists
55 (56.1%)	43 (43.9 %)	38 (38.8 %)

Table 6. *Loglinear model representing the adoption of interfunctional groups and DFX*

Variables in the model	Context : larger companies Objectives: life-cycle engineering paradigm Techniques: DFX, interfunctional groups Effects : -	
Effect	**Description**	**Parameter (log odds ratio)**
Two-way effects	DFX * life-cycle engineering interfunctional groups * life-cycle engineering larger companies * interfunctional groups	0.28 0.31 0.44
Three-way effects	DFX* life-cycle engineering * interfunctional groups	-0.11
Likelihood Ratio Chi Square = 2.76131 DF = 5 P = .737 Pearson Chi Square = 2.54550 DF = 5 P = .770		

data set still confirms these preliminary intuitions, and the objective of this subsection is to briefly discuss some of the elements regarding the relationship between organization and design, mainly from a descriptive point of view.

Concerning interfirm organization's effect upon design activities, a relevant classification has been made among (A) companies offering standard products through catalogues, (B) companies offering customized products and competing through tenders and (C) subcontractors which mostly compete within their customer companies' supply chains. The main criterion behind this partition is based upon the temporal relation between the product design phase and sales transactions, which tends to give a specific orientation to product development activities. Confining the present discussion to the qualitative phenomena emerged during field research, it has been found that type "A" companies design products first and market them later, so that the focus of product development efforts is the product itself. Type "B" companies operate upon an "Engineer-To-Order" basis; therefore, product design unfolds at the same time as transactions, since in general each tender is related to a specific design. For these companies the critical issue is to get a winning proposal out on time. Finally, type "C" companies' sales depend upon long-term and stable relationships with customer companies, which largely precede the design of each specific product. The focus of product development is the design process itself, since the main objective is to integrate it with the customer's one.

Concerning intrafirm organization, a most interesting element is the emergence of interfunctional groups in companies which formerly had a purely functional organization (see previous Table 4). The loglinear model related to this organizational form is reported in Table 7, and shows that its adoption is associated to the need of managing large numbers of designers on the one side, and to the technical problem of dealing with life-cycle engineering factors on the other side (this model does not also consider DFX tools, as in previous Table 6).

Interestingly enough, and despite their widespread adoption in the sample, the analysis of engineering teams' microstructure has shown that little rationale exists behind different organizational solutions adopted by companies. Hierarchical cluster analysis has allowed to classify quite distinctly three different profiles of interfunctional groups and as many profiles of team managers (details are here omitted for brevity); now, little or no association has been spotted among the presence of specific kinds of groups or team managers and other variables, both in context, objectives or effects. On the one side this may lead to hypothesize that companies have little knowledge of team engineering principles (in the sense of "designing" teams), so that the internal configuration of groups comes out at random, rather than as the result of a rational design activity. It anyway remains quite puzzling that no significant pattern explaining diversity among teams or team managers does emerge, at the least as the result of evolutionary mechanisms.

Because of this, it is possible to conjecture that interfunctional groups' role is something broader than simply that of a "technique" adopted in order to serve a specific technical objective. This idea comes from the observation that interfunctional teams (and, partially, interfirm organization), appear as statistically significant factors in about all of the loglinear models which represent the adoption of other techniques. The following Table 8 reports a list of techniques and some of their *modes d'emploi* (particularly for CAD systems) and shows the degree of association (if significant at 95% confidence level) with the presence of formal interfunctional groups, as reported by the loglinear model explaining the technique's adoption. The last column reports the same information, but concerning interfirm relationships. The extensive associations found between organizational factors and design support techniques therefore supports the idea that interfunctional groups may have the role of facilitating the adoption and implementation of techniques [13].

Table 7. *Loglinear model representing the adoption of formal interfunctional groups*

Variables in the model	Context : large company, belonging to the automotive industry, large number of designers, competition based upon tenders or strong supply-chain relationships Objectives: life-cycle engineering Techniques: interfunctional groups Effects : -	
Effect	**Description**	**Parameter (log odds ratio)**
Two-way effects	interfunctional groups * large number of designers interfunctional groups * life-cycle engineering	2.38 1.42
Three-way effects	none	-
Likelihood Ratio Chi Square = 81.7363 DF = 99 P = .8959		
Pearson Chi Square = 83.9920 DF = 99 P = .8594		

Table 8. *Summary of the effects of intra- and interfirm organizational factors upon adoption of techniques*

Technique	Effects of interfunctional teams (log odds ratio in loglinear model)	Effects of interfirm relationships
Anticipation of life-cycle factors in the product development process	0.36	-
CAD as a mean for communicating product data with other companies	0.13	-
CAD as a mean for communicating product data within the company	0.34	-
CAD for managing the product's engineering bill-of-materials	0.10	-
CAD for integrating engineering and manufact. bills-of-materials	0.16	-
Rapid prototyping techniques	-	0.46
Teleconferencing media	0.15	0.36
Groupware tools	0.35	-
ISO9001 standards	0.39	-
Project management techniques	2.62	-
Value analysis and engineering	0.63	0.50
Reliability engineering techniques	0.16	-
Quality function deployment	-	0.68

5. Conclusions

Theoretical reflections and preliminary empirical verification carried out in the context of the research described in this paper seem to support a view of engineering design in manufacturing companies as a very complex "melting pot" of different ingredients: managerial objectives, organizational factors, methodological techniques and technological support.

This paper has been primarily oriented to bringing into light and describing such complexity, while just a few attempts have been made to interpret it. Of course, a very interesting research question would actually be to unveil the mechanisms which lead to this pattern of relationships. The research will naturally continue in such direction, and a few hints have already been raised in the previous discussions: the development of a framework for understanding the mechanisms with which support techniques act (which may possibly be derived from the theoretical model in the first section of this paper), as well as further investigations upon the relationship between the organization and design activities. Nonetheless, a few conclusions may be drawn at this stage of the research, especially in relation to the topic of design productivity and to the methods for improving it.

First of all, it appears quite purposeless to consider the efficacy of individual elements concurring to engineering design processes; rather, a more comprehensive concept of productivity should be taken into account, which should embrace the entire design function of the firm, including its specific objectives and assets.

Moreover, the complexity of the phenomena which have been discussed suggests that a great deal of firm subjectivity does influence design and product development activities; despite the existence of common patterns which will be the object of further research, it appears that there is plenty of firm-specific "style" in incorporating techniques and using them for producing designs. The consequences of such a vision are quite important, if confirmed by future research; the two most direct ones can just be briefly mentioned in this section devoted to conclusions. On the side of engineering design research, efforts in developing complex systems or methods for supporting design should keep the flexibility of such techniques as one of the principal design objectives. The ideal goal would seem to be the development of an *ensemble* of "modular" techniques (i.e. methods, software tools, hardware) which companies may incorporate, adapt and combine with the maximum freedom as a function of firm-specific product development attitudes and past experience. Finally, concerning the management of innovation in product development practices, it appears that the sheer adoption of a technique is a quite meaningless event for a company, since the only important fact is its incorporation and amalgamation in the company's design function. In this sense, organizational factors seem to be at the heart of innovative processes, thereby constituting the most effective "knobs and handles" in the hands of management.

Acknowledgements

The research work upon which this paper is based has been funded by the OTEO initiative of the Chamber of Commerce of Torino. The author wishes to thank all of the colleagues which have cooperated with OTEO and the group of students who, with painstaking determination, have taken up much of the hard work connected to the surveying activity.

References

1. Dertouzos M L, Lester R K, Solow R M 1989 *Made in America. Regaining the Productive Edge*. MIT Press, Cambridge, MT
2. Clark K B, Fujimoto T 1991 *Product Development Performance*. Harvard Business School Press, Boston, MT
3. Molina A, Al-Ashaab A H, Ellis T, Young R, Bell R 1995 A review of Computer-aided simultaneous engineering systems. Research in Engineering Design 7:38-63
4. Vincente T 1990 *What Engineers know and how they know it*. John Hopkins University Press, Baltimore and London

5. Dosi G 1982 Technological paradigms and technological trajectories. Research Policy 11:147-162
6. OTEO Working Paper 1995 L'innovazione nella progettazione e sviluppo del prodotto industriale: un'analisi bibliografica e una proposta di sintesi per la ricerca empirica nel settore (in italian). Politecnico di Torino
7. Calderini M, Cantamessa M 1997 Innovation paths in product development: an empirical research. To appear in International Journal of Production Economics
8. OTEO Working Paper 1996 L'innovazione della progettazione e sviluppo prodotto nell'area torinese. Una ricerca applicata (in italian). Politecnico di Torino
9. Beatty C A, Gordon J R M 1988 Barriers to the implementation of CAD/CAM Systems. Sloan Management Review, Summer.
10. Holt K 1991 The impact of technology strategy on the engineering design process. Design studies 12:90-95
11. Liker J K, Fleischer M, Arnsdorf D 1992 Fulfilling the promises of CAD. Sloan management review, Spring
12. SPSS Inc. 1994 *Advanced Statistics Users' Manual, version 6.1*. SPSS Inc., Chicago IL
13. Calderini M, Cantamessa M, 1996 The dynamics of technological change and organisational structure: a study of small and medium firms. In Proceedings of the "Management and New Technologies" COST A3 Conference, Madrid, June

Negotiating Right Along: An Extended Case Study of the Social Activity of Engineering Design

Scott L. Minnemann and Steven R. Harrison
Xerox Palo Alto Research Center, 3333 Coyote Hill Road, Palo Alto, CA 94304

Abstract
We report on a 14 month project we undertook involving a small engineering design group (5-14) people within a large product program (up to 1000 people over a seven year period) in an even larger industrial setting ($16B annual, 100,00 employees). The research involved both observational work and technological intervention, the latter taking the form of supporting the designers' everyday work with recorded video services (recording and replay). Our findings then, reflect this two-pronged approach; we report both observations on design process and the utility of video to support design. Careful fieldwork and video analysis has resulted in a deeper understanding of the workings of design as a social activity, most notably in precisely how it is that personal and group histories and politics permeate every core of design work. Our use of video in the project has highlighted areas where multimedia tools are (or are potentially) useful for supporting the social aspects of designing and what needs to be done to more smoothly integrate these resources into designers' work.

1. Introduction
Recent times have seen a resurgence of interest in engineering design, fueled partly by growing pressures to shorten design cycles, reduce time to market, improve design quality and reduce overall design life-cycle costs. this interest has certainly produced more and better commercial tools for particular stages of designing; CAD tools are growing ever more powerful, with simulation and realistic 3-D rendering capabilities moving onto increasingly affordable platforms. Academic study of engineering design has also proceeded on a variety of fronts, with methodological developments and design tools dominating the research.

Still, what do we, the design research community, have to show for our efforts? Precious little, and certainly no phenomenal upsurge in design productivity (whatever *that* is). Why? The authors assert that our community has generally been crafting *engineering solutions for social problems.* For example, when faced with the growing challenge of multidisciplinary design teams, the design research community all to frequently is seen recommending a *lingua franca* or suggesting cumbersome decision-making aids.

A wide variety of studies over the last decade, in both engineering design and allied design fields, have highlighted the extent to which design must be thought of as a social activity, as the interactions of individuals within groups and the relations of groups to one another. Bucciarelli [1], for instance, used participant/observation methods in his studies of the workings of design activity within two very different engineering design groups. He writes:

> . . . *[the design] exists only in a collective sense. Its state is not in the possession of any one individual to describe or completely define, although participants have their own individual views, their own images and thoughts, their own sketches, lists, diagrams, analyses, precedents, pieces of hardware, and now spreadsheets which they construe as the design. This is the strong sense of 'design as a social process.'* [1, p. 161]

In a related design field, Cuff [2,3] has done extensive fieldwork focused on the negotiations between architects and clients. She argues against the prevailing myth of the architect as a heroic figure and for the adoption of the notion of architectural practice as negotiated social interaction:

> . . . *negotiation is part of the creative act of design, not only as a tool, but as a context for the emergence of buildings. That context is dynamic, evolving, and multi-faceted. It includes a unique constellation of participating individuals who interact in a manner and sequence which they themselves determine.* [2, pp. 23-24]

In order to understand and support design activity as it actually happens, in real-world settings, we must appreciate design practice as an intricate and complex technical *and* social undertaking. Focusing on design as a social activity results in a general orientation to the communications in and around design work. Design communications serve as the primary means by which the social work of designing is accomplished; by virtue of being available to other participants in the design effort, these communications are also available for design research. The method we employ in this research is aimed at unearthing and intervening in the detailed workings of everyday human activities. This research is not a study of group creativity; it focuses, not on the number and quality of the final results of a design activity, but rather on the processes that the participants engage in during the design activity. This focus permits a concentration on important initial questions about how interactions between participants are accomplished and what purposes those interactions serve.

Methods for understanding the activity of design as it occurs in natural settings are a problem. The claim is made that design research is in a pre-theory and exploratory stage [4], but that explanation is wearing thin and few researchers are using methods appropriate to that stage of understanding anyway. The work of Tang [5], Harrison and Minneman [6], and the exceptions noted above are making headway in developing and applying alternative methods for the study of designing.

Nonetheless, the findings of all of these researchers indicate that, when faced with a design situation involving a distributed[1] multidisciplinary design team, facilities must be provided that allow interested parties to negotiate to agreements.

1. Distributed, in this sense, includes both physical distribution (different places, possibly hundreds of miles, but hundreds of feet can also be a detriment to smooth collaboration) and temporal distribution (this, in the form of time zones) is often an entailment of physical separation, but we also mean to include scheduling difficulties and the possibility of participation via recorded media (e.g., minutes from a missed meeting, video).

In the heat of the moment, interest-relative negotiation is precisely what happens, regardless of what processes and tools are in place for a design team to use as resources. Politics, personal and group histories, and complex organizational interactions are the norm. As designers and educators, we must come to grips with these phenomena, to see them (and their ramifications) as fundamentally what design is about, not as deviations from a rational ideal.

We share the explicative side of this research direction with those engaged in particular branches of the social sciences, most notably the social anthropology and ethnomethodology communities [7]. The interventionist side of this goal is less common; it is a design orientation that is largely played down in traditional research. Walters [8] argues for "action research" in urban planning, an approach that attempts to integrate research and design; participatory design, too, is a step in this direction. In this vein, we have engaged in a series of projects aimed at improving design processes by enabling better communications within the design activity. Our efforts have focused on the use of video, both real-time and recorded, as a medium for design activity [6,9,10,11]. Video has proven an effective means for supporting the social and technical aspects of design activity. Video, then, is both a data collection tool for our observational studies and a technology we provide to our subjects as a tool to improve the work environment.

This paper then, will present a detailed account of a 14 month design effort involving, on the research side, both direct observation and intervention with video services. We will present some novel aspects of the social activity of design work that we have discovered in our *in situ* studies and return to the case study to analyze with and illustrate those crucial phenomena. Finally, we will present findings about our use of recorded video services in this setting, including implications for both the design activity itself and systems that might be (and have been) built to more directly support using video in designing.

2. An Introduction to the Case

This section describes, in some detail, the particulars of a case study we conducted within the confines of a large ($16B annual revenue, 100,000 employee) industrial concern. Like, we assert, all engineering projects of appreciable scale, the story is not simple, nor can it be made simple by omission without doing damage to the very things that make it a good case. With that complexity comes a degree of interconnectedness that makes a linear exposition of the detail difficult, but here we attempt to establish some categories—design project, major players, design and organizational context, research project—to help in getting everything out on the table. Many names have been changed to provide a modicum of anonymity for the designers we worked with.

2.1 "Data Collection"

The material collected about this design effort comes from a number of sources. As noted, the leader of this design project team agreed to work with members of a research team studying applications of video technologies in engineering design practice. Members of the research staff, the authors included, attended and videotaped numerous project meetings, laboratory tests, meetings with vendors, casual drawing reviews, etc. Members of the design team supplemented this data with videotape they shot of similar events when the research staff was not present. Over 400 hours of videotape and reams of documentation were collected during the 14 months of our involvement in their design work.

There were a number of aspects of the situation under observation that made the data particularly attractive for this research:

- a real group that had interdependence among members, differentiated expertise, reliable membership distinguishability, etc.,
- a task that the group collectively produced and that they were responsible for,
- some organizational context—the group had consequential interactions with outside entities and those interactions were collectively managed,
- a poorly defined task—part of the design task was to figure out the problem, and
- a poorly understood domain—it was imperative that design/development/research proceed simultaneously.

On the other hand, there were qualities of the situation that made it less than optimal for research and reporting results:

- it involved some proprietary subject material,
- the established work setting and relationships were often left unarticulated, and
- the diffuse activity meant only a small fraction was being captured on video.

While the decision to view naturalistic activity in a comparatively uncontrolled setting made the analysis work especially challenging, it appeared to be an appropriate means of gaining access to the pivotal issues in group design practice. Having the buy-in of the design team was an invaluable contribution—the interweaving of video as intervention and video for observation made for a unique opportunity.

2.2 The Deletions Study

The deletion project arose within a large-scale product design effort, dubbed Horchata. The product would include a xerographic subsystem, called an input-output terminal (IOT), that was supplied by another large-scale design project, Turret (actually, the story is slightly more complex, with another internal supplier between Turret and Horchata). The IOT was intended to fill the need of the Turret product (a high-volume copier) to print on plain paper (this will satisfy the bulk of the commercial reprographics market), but in the Horchata product (a very high-end printer) it would have to print on perforated paper (for simplicity, for the remainder of this paper, we will use this specific instance of the problem as our example), textured material, and other demanding paper stock. When used to print on perforated paper, the IOT did not function correctly, with printing faults, called deletions, occurring near the perforations. A deletion, in the reprographics and laser printing business, is a flaw in the printed page where, in a location where toner should have been deposited, there is none (or a noticeable lack).

The Horchata product program needed to modify the IOT from the Turret product to print without deletions on perforated paper, and it did not have individuals with the mechanical engineering or xerographic process expertise to make the modifications itself. Consequently, the program requested the formation of an engineering activity to modify the IOT. The activity was not formed within Turret, but within a separate technology development department. This spawned activity was the deletion project, and its charter was to design alterations to the IOT so that it could print without deletions on perforated paper.

As the project began, deletions were understood to arise during the transfer stage of the xerographic process. "Transfer" refers to the movement of a xerographic image, consisting of fine toner particles, from a photoreceptor to paper. Within an IOT (see Figure 1), a latent image is created on the photoreceptor by differential exposure to light (be it from a laser or traditional optics), toner is deposited on the charged portions of the photoreceptor and then the photoreceptor is moved to the transfer zone, where it is brought into contact with plain, unmarked paper. While the photoreceptor and paper are in contact (and moving together), the particles that constitute the image are electrostatically transferred from the photoreceptor to the paper. The paper, with the image deposited on it, is then separated from the photoreceptor belt; the belt is cleaned and recycles back through the machine to receive another image. The toner image, electrostatically adhered to the paper, is now heat fused to the paper and the page is delivered to the user, complete.

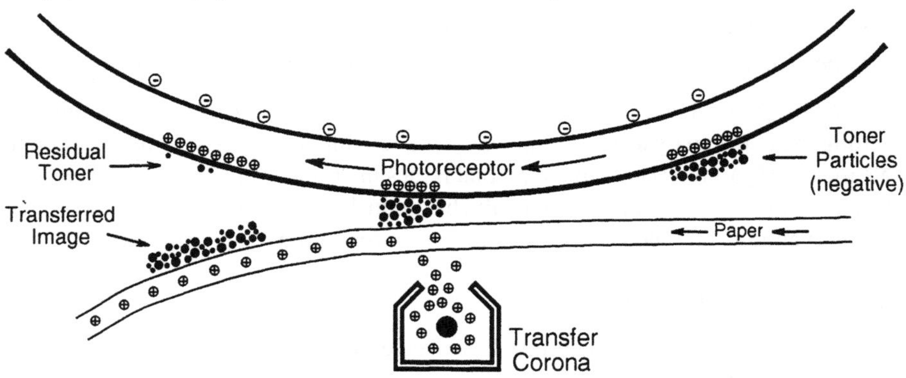

Figure 1: Transfer Zone (in cross section) of an IOT

When the transfer process works properly, essentially all of the particles that constitute the image are deposited, unmolested, onto the paper. However, when particles remain on the photoreceptor, rather than transferring to the paper, corresponding portions of the image are "deleted" from the paper. In the case of perforated paper, deletions tend to occur along the perforation line. The perforation deforms the paper, so that it does not lie flat on the photoreceptor, leaving a substantial gap between the paper surface along the perforation line and the photoreceptor (see Figure 2). If the gap is too great, electrostatic forces do not transfer the toner particles, leaving them on the photoreceptor, and causing deletions to occur.

There is not a widely accepted formal description of the deletion problem, nor is there, within the industry, an exhaustive battery of tests that can be run to identify or quantify the problem. Part of this is due to the fact that the single term actually refers to a multitude of problems. Deletions may result from a number of other causes, including tenting of the paper around stray dirt particles, from paper curl in areas where the opposite side is black (from differential shrinkage) on two-sided copies, or from using other odd paper stock (e.g., fake parchment).

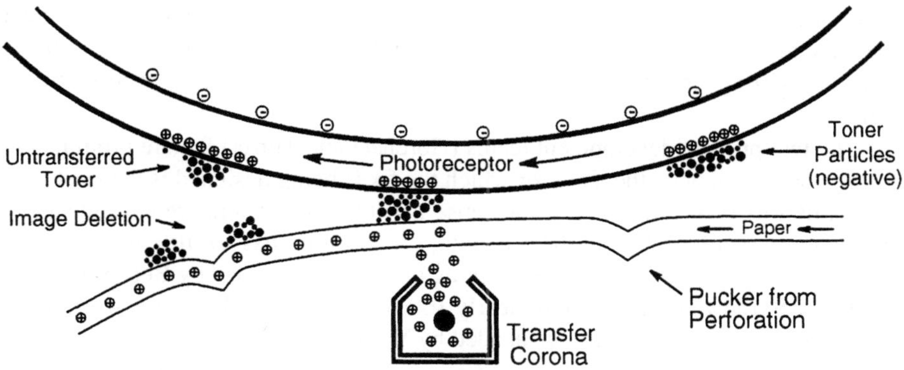

Figure 2: Exaggerated view of problem area with uneven papers.

Based on their existing understanding of deletions, the technology department studied the Horchata deletion problem and proposed a project to address it. The study considered two major approaches: first, forcing the paper to contact the photoreceptor, and second, causing the particles to jump across the small air gap from the photoreceptor to the paper. The former was judged to be the more promising approach, and the project was accordingly planned to develop a solution that would cause the paper to contact the photoreceptor. The latter, causing the particles to jump an air gap, is used by competing products and can exhibit blur problems; the possibility of blur, coupled with the change in process technology that this approach would require, doomed this option.

The initial study of the deletion problem took place in August and September several years ago; in October, the plan was reviewed and accepted by the Horchata project. In November, Lester, a manager in the technology organization, was assigned half-time to the project and it moved into its first active phase.

2.3 The Major Players

Lester manages a small group in a large development and production facility in a city in New York state. The charter of his group is unusual, they do contract engineering, using "state-of-the-art" methods, for other organizations within the corporation. As such, they both exist and they don't; a large portion of their money comes from organizations that fund them, so they don't necessarily have people "at the ready" when they are called upon to do a job. The implications of this will repeatedly show up in this case.

As the project progresses, Lester assembles a team of physicists (experts in the xerographic process), engineers (what we might call engineering designers), draftsmen (officially called "designers"), and technicians. This process of assembling teams and defining roles will be covered in a later section, suffice it to say that the process is a fluid one—seldom could the team be precisely defined with an organizational chart.

Lester's boss, Roland, runs an advanced technology development group within an Architecture of Systems and Products (ASP) division. He is broadly interested in improving the design process at the corporation; this is evident in methodological work that he encourages people under him to employ in their projects, and also shows up in his willingness to involve the video services research team in this critical project.

The two product development teams (Horchata and Turret) feature prominently in this story. Curiously, the team for which the fix is being designed is the least visible; while this is unsurprising in the sense that these designers are not physically proximate, it is a bit odd that the receiving organization has so little involvement. The local product program, the one working on Turret, is headed by Bob Winter and numbers 250 (down from a peak of nearly 1000). The remote designers, working on the Horchata product program, are headed by Vern Cole and number about 300.

The research team was made up of architects (design researchers, all), a video producer, a mechanical designer/engineer, and a design research consultant. As noted above, we, the research team, became involved because Roland saw potential in our research approach and agenda. We proposed to support Lester and his design team with recorded video services, essentially simulating an elaborate semi-automatic recording, indexing, and editing facility. To this end, we arranged to be present for open-ended interviews, major meetings, and other significant project events; we also enrolled the design team in the agenda of recording much of their everyday design work. We also repeatedly provided the design team with interventions of various forms—produced videotapes for particular occasions, better equipment for troublesome recording situations, a videodisc browser for project history, etc.

2.4 The Design Context

In this case study, Lester's organization is being called upon as an emergency response unit for a product development effort in trouble. Although it is rare that a group of outsiders would be more appropriate for addressing such difficulties than the original designers, in a corporation this large, there are many reasons that might crop up to make this the case. Examples include the dissolution (usually by reassignment) of the original design team before the problem is discovered, management's belief that a fresh approach (using different people) is necessary, and the discovery of the problem by another product development effort hoping to re-use a component in another product.

This last case (compounded by a measure of the first) is what has happened to Lester and his cohorts. Migrating the print and development engine from one reprographic product, a copier, to another, a printer, had revealed a performance problem with deletions. While the copier team, who had done the bulk of the engine's development, knew of the problem, it was considered a serious nuisance for them. For the printer folks, it is a show-stopper—their customers have a set of demands that don't perfectly align with those of the copier market. Luckily, their introduction date trails Turret by nearly a year, so they have more time to work this problem.

One complication in this effort is the geographic location of the various players. The research team is located in Northern California, the Horchata printer product team in Southern California, and the Turret copier development team and Lester's group are in different buildings in Central New York. When we first started working with Lester, he had had little direct contact with Vern on the specifics of the performance problem, and there was little understanding of the severity of the problem or the latitude they were going to have in coming up with a solution. What is clear from the outset is that the problem that Lester's group is faced with is very important to the organization he's working for; they need to find a solution or the printer division's customers will likely be unhappy.

2.5 The Research Project

The research team, then, was to support the deletion project with a variety of video services. Along the way, we were privy to the workings of an extended design effort. The support that we provided included: recording project status meetings; individual interviews on the status of the design alternatives; capturing activities in test labs; interviews of the Turret technical program manager; and planning for video communication between the deletion project and the Turret staff about deletions. Over the course of this study, various members of the research staff collaborated in the production of several video documents that the designers used to communicate the state of their design effort to others in the corporation. In addition to directly providing video support, we delivered a multimedia videodisc interface to the deletion project's recorded history. This interface permitted experiments with access to video records by members of the deletions project; for example, looking at how recordings were used as illustory material when bringing new members up to speed.

The recording in this study was not a neutral data-collection activity, the recorded segments were subject to re-use in any setting where it was deemed fit. The responsibility for deciding on segments to be included in videotapes was split between various members of the research team and Lester. Project participants had the option of erasing particular footage as it was being sent off for indexing, but signing up to work with Lester on this effort meant, in part, that one's activity was going to be public (this is not unlike joining a design consultancy that embraces a public style of working, but it was unusual within the corporation's product development organization, and it was a small pocket of activity that had this understanding).

One specific activity we will be examining in this paper is taken from an early design meeting that Lester has called in order to get a better handle on the nature of the problem that he has been charged with solving. He has invited everybody he knows who might have an opinion on the topic. A salient aspect of the organization that Lester works for is that they are interested in developing and using state-of-the-art processes, this includes why we are working with them and systematic methods they employ for doing their project work. One method that their group was pushing in the days that we were working with them was the Pugh method for concept selection.

3. The Social Activity of Designing

This work is built on a few very specific (and possibly controversial, when taken as far as we do) premises about design: design is the creation of something new that fits with reality; it is usually carried out through interaction with others; it is based on experience; and it is unrepeatable, since the practice of designing changes the participants.

- *Design is the act of making something new that fits with reality* [12,13]. Design is not unbridled imagination, but a fit with the world and its messy and contradictory conflicts. This definition is inclusive—it takes in all of the influences that shape the final result and all the steps necessary to generate the result. Design is not an abstract result or style, but a generative process that takes idea into reality.
- *Design is the process of socially constructing a technical reality* [14]. It is not the work of individual designers, but the interplay of individual work and the relations between designers. This social construction is one of the most visible of design process phenomena.
- *Design is experiential.* It is not described adequately in external representations; it exists in the moment; it is taught by doing. The act of designing changes the designer, making it impossible for the same design to truly be done twice [15]. Designers reflect and project.[2] They reflect on the experience of design, and they project an imagined existence as a real place onto their designs. Stories are generated to flesh out the reality of these projections; the stories are shared and become the basis of the description.
- *The result is to create a shared understanding among the participants.* The shaping of the understanding occurs as a result of imagination, conversation, experience, power relations, personal dynamics, persuasion, and personal insight.

This is not a cognitive or problem-solving view of design. Although problem solving is one kind of design activity, it is not the entire story. The problem with using a problem-solving model is that everything that is a "solution" requires a problem to motivate it, but motivation in design is not a clear-cut rational process. If we ask, "Why does this look or work a certain way?", we want the true history of it, not a post-facto rationale.

What sort of communications occur in design? Design communication is the vehicle for developing a shared understanding among the various players. The degree of *ambiguity* in the communications is the key to providing the "communications space" for a common understanding to develop through exploration and explication. Ambiguous communications provide an opportunity for designers to project and reflect—breathing room from rational concerns. Designers project a story onto suggestive fragments to make a whole, creating the shared understanding.

Just as this is not a problem-solving view of design, neither is it an information theoretic view of communication. Although the content of design communications are describable in information terms, the communications are subtle, mutually developed among the parties, and highly contextual.

2. This is an extension of Don Schön's [16] notion of the reflective practitioner.

The phenomena of design communications are the most visible and accessible parts of the design process to those on the outside. As an outside observer, it is easy to confuse the communications with the whole story of design activity. The communications can engender and demonstrate activity while not being the activity itself. For example, someone may ask the question, "How wide is it?", which is answered by "About 4 inches". The answer probably required some calculation and an understanding of the complex context in which it was asked. But the common "truth" of the design activity, the one that is shared by the group, is the question *and* the answer. Thus, it is more than a data collection and reduction technique, it is the greatest common denominator that a group defines itself with [14].

4. Our Example Again, More Closely

Returning to our case study, let us now take a closer look at the designers' activity for instances and ramifications of the phenomena we have brought up in the previous section. These points are but a subset of the many that we could have discussed. The points chosen are intended to give readers a flavor for the breadth of the findings. We cover them in widely differing amounts of detail in this publication—some careful, to give the reader a feel for the kinds of evidence that are present in the data, some are more suggestive, in the interests of brevity.

4.1 Negotiation

In this design effort, everything was negotiable. Lester negotiated with Turret and Horchata to get initial resources, the parameters of the deletion problem and solution were argued through, the numerous individuals were enrolled in the design work through give and take, the role of the research team was constantly redefined. Process, artifact, and relationships were all sorted out through interest-relative negotiations.

Although these negotiations were often carefully handled, other instances were put quite contentiously. Here is an excerpt where one experienced designer is criticizing the use of a particular decision-making technique:

Tape excerpt 1: MDP.12.10.3

DK (engineer):	this procedure, for selection, seems to me to be appropriate when you have a range of possibilities, some or all of which will satisfy your minimum requirements -- you are trying to pick the best, most cost effective, most schedule effective...
LL (manager):	right
DK:	...one among them. Unless things have changed radically since the last time I looked at this problem, we are instead confronted with a range of possibilities it's quite possible none of which will satisfy the requirements...
LL:	Ok, but...

DK:	...It's not obvious to me that this is the appropriate tool, but, since I don't have an alternative, I will suspend judgement.
LL:	L-l-let me do this, then, l-l-let's let's keep that in mind and and uhh I'm I'm going to make a note here that that we will reconsider the Pugh method after we've gone through the concepts and then we'll see if you feel any differently.

Contentious again, this time from an interview, Lester talks about the relationship between his group and the Turret team currently assigned to solve the same (less severe) problem. Lester grows increasingly frustrated over the course of the project about his access to resources, and is embroiled in constant battles to get the rare prototype machines, technicians with experience in keeping the prototypes running, and, always, money. This excerpt is from a memo videotape that was requested by the Turret project manager and is shown at a Turret team project meeting at a crucial point in these negotiations.

<u>Tape excerpt 2: MDP.3.2.1</u>

LL (manager):	The stand I'm taking is that, if Turret is going to work this problem, then they oughta work this problem and leave me alone. Ok? ...take back your machines, take back your people, take back your problem, Ok? ...and I'll do what I can, focused on Vern Cole. If...and, and then Turret you can work your own problem....if, on the other hand, you're going to work the problem, and then you're gonna you're gonna compete with me...I'm having difficulty with that...you know, because you're the customer, see? ...if you're working on the problem and I'm working on the problem, I'm not comfortable that we're going to look at all the alternatives with an objective eye, Ok? ...my personal opinion is that Turret doesn't have a prayer of doing something positive with the approach they're taking...that's just where I am...and my own personal opinion with my own integration undefined...Ok? If they have people and headcount and money to spend on this project, send it over, I can use it.

Oftentimes, engineers will admit to negotiating about process and reporting relationships, but hold what they consider the design itself as somehow sacred, as independent of those negotiations. This does not, in our experience, stand up to much scrutiny. In this case study, for example, the nature of the deletion problem itself was under constant definition—to what extent it could be remedied, for what papers, with what impact on other aspects of machine performance. Even more obvious was the extent to which particular solutions would require organizational work. Real estate in the machine was precious, with a major frame member close nearby. Displacing other components with pieces of the solution seemed very likely, and would require complex negotiations with those who had developed that part of the machine. Furthermore, any solution that protruded into a location that inhibited jam-clearing or common service maneuvers was going to be unpopular downstream. It is difficult to fully ascertain the extent to which this colored the design work, but it was significant; we'll touch on this in the next section.

4.2 Politics and History

Project and organizational history and politics absolutely permeate this effort. We came to appreciate many of these facets over the course of our research project, but there doubtlessly many more that were at play in a product development program of this size. Tensions abound, on many levels, and surface at many levels in the design and the design effort. The corporation was none too quick in seeing the potential disruption that would result from this performance flaw, and many laid the blame for the late discovery of the problem at the feet of the upper management that had eliminated corporate-wide research programs into many aspects of the xerographic process. A decision had been made that xerography was well-understood and that product programs needing process expertise would come to the ASP for it.

Xerography and materials research continued, but now was essentially sponsored by product programs. This encouraged the use of product programs as covers for risky advanced research, and resulted in numerous late-cancelled product programs. In the case of Turret and Horchata, their product programs served as a vehicle for the introduction of an important new flexible-belt photoreceptor material which, while having been thoroughly tested itself, had never been integrated into a machine of this scale. Xerography was understood, but best within a set of confines defined by existing machines—Turret/Horchata were outside those bounds, and some people were unsurprised by the difficulties that were cropping up.

Particular members of Lester's design team and his set of volunteer advisors were completely wrapped up in this bit of corporate history and its entailments. DK, for instance, expressed harsh assessments of the prospects for finding a quick fix to these systemic problems; his opinions had to be (and were) filtered by the other project participants—everyone was aware that it was possible that a major tactical error had been made, but that it didn't change the current situation.

The deletion problem generated its own internal history very quickly. When we attended an early brainstorming meeting in December, the attendees were bemoaning that there couldn't possibly be much new to be added, since no real (technical) work had been done since an earlier meeting when the problem had first been uncovered, in August. The crucial difference was that there was now an organizational context for additional work to be done; state needed to be shared, and promising directions needed to be settled on.

4.3 Enrollment and Membership

Who was working on the problem was an ongoing issue in the unfolding of this effort. Lester had limited resources available to him, so he was constantly on the lookout for volunteers and people with "cycles to spare". On the flip side, Lester was seen by many others as a potential source for funding and a sort of certification of approval/relevance. For instance, two of the concepts that Lester ends up pursuing in this work have their genesis in a nearby skunkworks that is always looking for ways of establishing the value to the corporation. Lester's efforts are seen as a way to get air time for ideas and perhaps capital for going beyond bench prototyping and into real machine builds.

The Deletions Project identifies itself, in part, by their association with a technical difficulty, in part by physical location, in part by managerial ties, and even in part by their association with us, the research team. They are, in the truest sense of the phrase, a multidisciplinary team—a clone of no one of them would be capable to tackling this problem. At times, despite the detrimental effects of their being in competition with the portion of the Turret team that is working on the copier's version of the problem, that competition results in a strong bond among the core team members.

4.4 Ambiguity in Communication

Ambiguity, in spoken language, text, sketches, gestures, and silence, is an important element in the designers' repertoire. Ambiguity is artfully employed to pull off the negotiations, indicate future process, preserve design latitude, and avoid unnecessary conflict. The ambiguity arises, not only from explicit communications, but also from those things left unsaid. Particular individuals and groups will have their own views of negotiated positions. These differing perspectives are not necessarily undesired—upon discovering discrepancies, participants discover new things while reconciling their differences.

4.5 The Power of Stuff

Over the many months and myriad meetings we attended or viewed, one constant was present: those with something to show got a chance to tell. This intriguing finding has repeatedly surfaced in our studies of real-world designing, and involves the participants' use of mundane text, video, and graphic representations. By mundane, we do not mean to suggest that they are boring, rather that, as an initial definition, they exhibit qualities of being practical, often temporary, and devoid of any extraordinary semiotics. Participants use these representations both as a means to tie their minute-to-minute activity to the overarching design effort, and as a means to command and structure stretches of interaction. For example, when a group puts together and fills in a specifications list, it serves to organize a stretch of design activity—all of the interaction about the entries on the list is seen as relevant to the task at hand—and has the additional value of provide a future framework for talking about the artifact. Entries in tables and on lists are valuable allies in arguing particular positions; they represent acknowledged ways to refer to salient qualities of an emerging artifact.

4.6 The Failure of Methods

Going into further detail about the example from the previous paragraph, at one point, early in this project, a group meeting is held using the Pugh selection method. As illustrated in Tape Excerpt 1, there is some trepidation about the use of this technique, but Lester has been trained in its use and is encouraged, as part of Roland's organization, to promote its inclusion in their design process whenever possible.

What transpires during the course of this meeting, although characterized by some as a disaster, is a useful and necessary discussion. There is no way, at the outset of the meeting, that a selection matrix could have been decided upon, let alone populated for the competing designs. At the end of the meeting, the participants are nearly at a point where they can use a selection method. However, when the technique is used, it produces the "wrong" result—the meeting attendees know what outcome they've decided on, and the method doesn't agree. The following time is spent changing aspects of the selection matrix to agree with the decisions they had already made.

Over and over, we have observed that design methods might best be thought of as frameworks for participation, not as process blueprints with a specified result. In the Deletion Study, the Pugh method is used to have a useful discussion, one that helps to tame the designers' unruly impressions of the design options.

4.7 The Distrust of Computing

In this project, there was a distinct lack of confidence in any kind of simulation. It was not the case that models of forces needed to accomplish particular paper deformations, for instance, didn't exist, in fact, a great deal of effort had been expended over the years in sophisticated models of many aspects of the xerographic process. It was more that there were thought to be too many steps from the initial assumptions to applicability in the complex environment of the transfer zone.

5. What About the Video?

Remember here, that video played two important roles for us: it was a window into the processes of the designers in our study, and it was fodder for a wide variety of video-based interventions into that design team's functioning. The previous section was derived from direct observations and from the use of the video in its first form. This section highlights some lessons about video in the second sense and includes suggestion about how the role of video in design activity might be exploited.

Further, remember that there are two consumers of our work with video: Lester and the other designers on the Horchata deletions work, and Roland and his organization concerned with improving design practices within the corporation (this second consumer is not unlike the design research community, in terms of deliverables).

5.1 Customer #1: Lester and his Team

Our primary customer in this study is Lester Trevino and the rest of the staff of the Deletion Project. Our deliverable to them consists of the services of video connection: recording, indexing, accessing, editing, distributing and playing video. In some cases we provide all levels of production, in others we provide capabilities and the members of the Deletion Project did their own production. But in all cases our emphasis is on providing the project with video support for its activities (while sensitizing project members to the potential that video offers).

Early in the project, our efforts primarily consisted of collecting background video material that we used as setup in subsequent tapes. In other words, we were getting up to speed on the project and were getting the various project participants to tell us (in open-ended interviews) what they knew of the people, the politics, the technical options, and the problem. This period lasted for approximately a month, although it would also include several months of vague setup work that might have been useful to have recorded (e.g., had we been involved at the time, the problem discovery and initial discussions in August would have provided insight into attitudes that were cemented by the time we became attached to the project).

As the effort began to ramp up, the role of video recordings began to expand. We began to serve as a communication conduit between organizations, most notably between the Horchata customer on the west coast and the design team in New York. Video was also used to communicate the problem to consultants, and to present material for comment to keep Lester's expert advisor team engaged. Additional video capture opportunities are explored, including staff meetings, one-on-one meetings in Lester's office and lab procedures. We also began identifying choice bits for inclusion on a random-access videodisk system to permit project participants to craft their own presentations.

As time progressed, the research team was repeatedly called upon to deliver edited videotapes for particular purposes. These included reports to Roland about how the project was progressing, updates to Vern Cole, and negotiations with the sister team in the Turret product program. In making these tapes, the mechanical engineer and video producer on the research team collaborated to produce viable narratives that accurately reflected the state of the design work. Although these narratives might have been expected to instantly reveal the "outsider" role of their authors, we found this not to be the case—our careful monitoring of the team's process recordings and occasional conversations with project participants was resulting our having a decidedly "insider" perspective on the project work.

Over the course of our work with this development team, the project participants, Lester and Bob Winter, in particular, came to know video as a distinctly political medium. In many senses, the video medium could be employed to communicate things that couldn't be said in other ways. Those on video can speak to the camera (or to the interviewer, if that's the format) directly about the anticipated viewers, objectifying them in a way that would be offensive in person. For example, the tape shown to Winter's (Turret) team contains frank assessments of that team's progress that Lester would have been unwilling to deliver in person.

The recorded video had some pragmatic uses, too. Lester is able, in some sense, to be in multiple places at the same time. Although the update video from which Tape Excerpt 2 is taken was requested by the recipients, Lester's presence at that meeting wouldn't have been possible anyway—he had a schedule conflict.

5.2 Customer #2: Roland and his "Improve Design" Agenda

Our second customer in this study is Roland Gomez, and other managers in ASP. Our deliverable to them is a description of video uses that show promise for improving design practice within ASP. Further, if possible we should project video uses into design practice in order to explore specific improvements, such as schedule length and predictability, or technology readiness.

The description of video uses, if it was to be successful, must take a form that help Roland and other engineering managers to take action. We believed that a written discussion could not serve that role. A written characterization of the uses of video is too remote from the experience of engineering activities to be credible to active engineering managers—in terms of the projects in the experience of engineering managers, such a discussion is taking place in a foreign culture. Similarly, the engineering managers will probably not be able to make use of our written projections of video uses into specific improvement in design practices.

The project itself was our most powerful demonstration of the potential for video use in design activity, however, we tried, at various times throughout the effort, to distill our impressions about the uses and potentials for multimedia in design practice. Lester was centrally involved in many of those discussions and produced the following list in a memo late in the project:

- Brainstorming meetings where a great deal of information is being exchanged and the meeting members are deep in thought about concepts and applications therefore unable to fully appreciate all that is being said. Here, all participants would review the video to glean otherwise lost opportunities.
- Meetings where concepts, plans, etc. are discussed but the "ideal" participants are not available at the meeting place and time. The absentees could then add their input, making the video record more valuable.
- Communications where maximum information density is desired such as progress reports to management and other organizations. Also, this form of communication could be used for top down information dissemination from senior management and could be used for orientation of new members to a team including items such as program activities and specifications.
- Review of meetings and other interactions to be used as personal development tools, study of group dynamics, meeting facilitation, team building, interactive skills assessment, etc.
- Presentation of information which is difficult to deliver or receive in person. This video might be used to challenge a group or deliver information that should not be diluted with listener objections and reactions. People who are watching video are not preoccupied with how they will direct the conversation or respond to criticism, they are therefore able to listen more carefully.
- Records of tasks that would be archived for post mortems, patent support, etc.

In the same memo, Lester claims that he found the video very useful in understanding the customer requirements as put forth by Vern Cole. His assessment of the video he sent to the Turret team was mixed: it produced the result of getting a prototype machine and challenged them to produce a better solution, but it set up damaging tensions between the teams, encouraging them to oversell their solution.

In hindsight, the Turret solution has indeed proven very problematic, and some of the more advanced concepts that were being developed by members of Lester's team are showing up in current products and are being further refined for inclusion in future products.

6. Further Discussion and Conclusion

This work focuses on design as an *activity*, a human practice, rather than on design as the artifacts being produced or values as represented by artifacts. This choice of focusing on actual design activity has the effect of elevating *whatever the designers do* to be the subject of the inquiry. We begin without an orientation to a prescriptive notion of what should be happening in the design activity (cf., the study of Hales [17], which regarded the actual activity as a flawed enactment of the normative view of Pahl and Beitz [18]), or a received theory of cognitive process (cf., the work of Akin [19, which analyzed design activity within an information processing framework).

The group engineering design work in this case can effectively be characterized as *negotiation for the sake of understanding*. The subject of those negotiations is wide open; nothing is off limits. Groups and individuals negotiate responsibility, process, artifact, properties of materials, interpretations of analysis—virtually everything. On the other hand, unlike the dictionary definition of negotiation (to treaty with another or others in order to come to terms or reach an agreement [20]), these interactions generally lack closure—participants seldom settle on a particular value or decision at the close of a negotiation (in fact, participants seldom mark the closure of a negotiation at all). Instead, participants use negotiation to come to a better understanding of each other's positions; the process is also the outcome.

The design in this case study is inseparable from the personal and group histories and the politics of the situation. This is true in the strong sense, not only that the interactions of the participants is rooted in these aspects, but that the entire design enterprise is so finely intertwined with these factors that the project wouldn't make sense without the organizational context. Contrary to popular belief, the technical work of designing is as thoroughly embroiled in the politics as the funding or group membership.

Multimedia resources hold considerable potential for supporting the social activity of design. We found the properties of video, in particular, well suited to the salient aspects of design communications. A number of improvements in design processes are possible, and they contribute to what we, somewhat jokingly, call "the tyranny of collocation" that designers are often faced with in real-world projects. Going beyond the benefits of real-time connection that we've demonstrated in previous projects, this research shows how recorded video can also benefit designers and contribute to scheduling and geographic freedom.

The design tool implications of this work are a shift in focus onto appropriate communication systems for groups. In the development of such systems, we must make certain that designers are able to reap the potential benefits—capture and indexing of process recordings must be kept simple and unobtrusive; editing and playback must be easy enough that users will utilize the resource and not be forced to spend undue effort or time. Technologically, the particular video service we provided to Lester's group (i.e., analog video and manual indexing) would not be practical in a non-research setting; few groups would be able to justify the efforts of a group historian and video indexer. On the other hand, systems that support real-time indexing and random-access replay will go a long way in overcoming the tedium of working with these media. Minneman et al. [11] describes a system that was derived directly from the lessons of this study, and Moran et al. [21] reports on our experiences in using that tool in a setting not far removed from designing.

This project also serves as an example for a proactive way of performing design research in practice—a way of intervening in design practice, of watching (and accounting for) the effects of those changes, and of planning additional interventions.

7. Acknowledgements

We would like to thank our colleagues Bob Stults and Karon Weber who served on the research team with us and contributed much to the success of this project. Many other researchers at PARC have also contributed to our understanding of this work, and many helped with the mechanics, as well. Thanks to Larry Bucciarelli for numerous constructive conversations during the course of the project itself. The design team also gets our undying gratitude—without their cooperation and patience, this project never would have happened.

8. References

1. Bucciarelli, Louis L., "An ethnographic perspective on engineering design", *Design Studies*, Vol. 9, No. 3, Fall 1988, pp. 159-168.

2. Cuff, Dana Charlene, *Negotiating Architecture: A Study of Architects and Clients in Design Practice*, Ph.D. Dissertation, University of California, Berkeley, 1982.

3. Cuff, Dana, *Architecture: The Story of Practice*, Cambridge: MIT Press, 1991.

4. Finger, Susan, and John R. Dixon, "A Review of Research in Mechanical Engineering Design. Part I: Descriptive, Prescriptive and Computer-Based Models of Design Process", *Research in Engineering Design*, Vol. 1, No. 1, 1989, pp. 51-67.

5. Tang, John C., "Findings from Observational Studies of Collaborative Work", *International Journal of Man-Machine Studies*, Vol. 34, No. 2, 1991, pp. 143-160.

6. Harrison, Steve R. and Scott L. Minneman, "Tools, Communication, and the Nature of Design", *Proceedings of ICED '93*, August 1992.

7. Heritage, John, *Garfinkel and Ethnomethodology*, Cambridge: Polity Press, 1984.

8. Walters, Roger J., "Informed, well-ordered and reflective: design inquiry as action research", *Design Studies*, Vol. 7, No. 1, 1986, pp. 2-13.

9. Weber, K., and S. L. Minneman, *The Office Design Project* (a videotape), Palo Alto, CA: Xerox Corporation, 1988.

10. Stults, Robert., "Experimental Uses of Video to Support Design Activities", Xerox PARC Technical Report, SSL-89-19, December 1988.

11. Minneman, Scott, Steve Harrison, B. Janssen, G. Kurtenbach, T.P. Moran, I. Smith and B. vanMelle, "A confederation of tools for capturing and accessing collaborative activity", *Proceedings of Multimedia '95*, November 1995.

12. Caplan, R., *By Design*, New York: St. Martin's Press, 1982.

13. Stults, R., *Shoptalk 1 - Representing the Process of Design* (a videotape), Palo Alto, CA: Xerox Corporation, 1985.

14. Minneman, Scott L., *The Social Construction of a Technical Reality: empirical studies of group engineering design practice*, Ph.D. Dissertation, Stanford University, 1991.

15. Stults, R., *The Media Space*, Palo Alto, CA: Xerox Corporation, 1986.

16. Schön, D. *The Reflective Practitioner*. New York: Basic Books, 1983.

17. Hales, Crispin, *Analysis of the Engineering Design Process in an Industrial Context*, Doctoral Dissertation, University of Cambridge, 1987.

18. Pahl, G. and W. Beitz, *Engineering Design*, London: The Design Council, 1984.

19. Akin, Ömer, "How do architects design?", *Proceedings of IFIP WG 5.2 Conference on Artificial Intelligence and Pattern Recognition in Computer Aided Design*, J. Latombe (ed.), North-Holland Publishing Company, 1978.

20. Morris, William (ed.), *The American Heritage Dictionary of the English Language*, Boston: Houghton Mifflin Company, 1976.

21. Moran, Thomas P., Patrick Chiu, Steve Harrison, Gordon Kurtenbach, Scott Minneman, and Bill vanMelle, "Evolutionary Engagement in an Ongoing Collaborative Work Practice: A Case Study", *Proceedings of CSCW '96*, November 1996.

Influences on Design Productivity - Empirical Investigations of Group Design Processes in Industry

Eckart Frankenberger* and Petra Badke-Schaub**
*Maschinenelemente und Konstruktionslehre, TH Darmstadt,
Magdalenenstraße 4, D-64289 Darmstadt, Germany
**Lehrstuhl für Psychologie II, Universität Bamberg,
Markusplatz 3, D- 96045 Bamberg, Germany

Abstract

In the research project 'Teamwork in Engineering Design Practice[1]' engineers and psychologists are investigating engineering design processes of teams in industry in order to get a deeper understanding of the interdependencies in design practice influencing design productivity. The overall aim of this project is to outline a model of group design processes as a basis for further development of systematic design with special emphasis on teamwork.

In this paper, some problems in design practice and the aims of this research project are introduced in the first two chapters. In the third paragraph, the approach and the procedure of the investigation are discussed. The content of the next three chapters (4, 5, 6) is the description of the methods used for compiling the relevant data of the design process and the external conditions, the task, the prerequisites of the individual and the group. The evaluation of the data and the substantiation of a model of collaborative design work is explained in the seventh chapter. First results concerning design productivity are presented in Chapter Eight. The last chapter discusses the actual state of the project.

1. Introduction

Successful companies have to meet the general demand to develop products of higher quality at lower costs in even less time. Thus, the 'productivity' of the product development process can be described by the characteristics 'quality', 'costs' and 'time'. These demands as well as the increasing complexity of design processes require a more parallel cycle of work in product development, as opposed to the traditional mainly sequential cycle. Consequently, engineering designers are collaborating more and more in teams accross both departmental and company borders [28]. In this situation, engineering designers are struggling not

[1] This research project is sponsored by the German Research Council (Deutsche Forschungsgemeinschaft DFG)

primarily with technical problems, but rather with difficulties related to their environment (e.g., effective organization) and to their colleagues, as surveys concerning the problems of engineering designers in industry have shown [8, 18].

Obviously, design productivity is very much related to the effectiveness and efficiency of cooperative work in groups. New demands on communication and cooperation raise important questions, such as how to organize teamwork effectively in complex working conditions, how to lead and to communicate within a team, or which individual characteristics are important for a good team member.

Regarding these questions the aim of research should be to identify the determining factors of teamwork in complex working environments. However, there are hardly any scientific findings about how team members with different individual prerequisites in different teams under different external conditions are solving different types of tasks in optimized collaboration. Unfortunately, the knowledge on the conditions for effective and successful problem-solving behaviour is based essentially on investigations in laboratory situations.

This gives rise to the question of how to create effective teamwork in engineering design involving both engineering and psychological aspects. In the research project 'Teamwork in Engineering Design Practice', engineers and psychologists are investigating engineering design processes of teams in industry. This research is based on the empirical research made so far on engineering design by engineers, ergonomic scientists and psychologists. Figure 1 shows an overview of the investigations of single designers and design teams in a surrounding laboratory and in design practice.

approach	external conditions	 single subjects	 group
design-scientic	laboratory	Upmeyer [43], Beitz & Luczak [7] . . .	Radcliffe & Slattery [37] . . .
	industrial practice	Beitz [6], Frieling & Derisawi-Fard [26] . . .	Hales [29], Blessing [9], Minnemann & Leifer [33], Radcliffe & Slattery [37] . . .
design-scientific + psychological	laboratory	Ullmann [42], Rothe [38], Ehrlenspiel & Dylla [17], Fricke & Pahl [24], Dörner et al. [14], Christiaans et al. [11]...	Badke-Schaub [3], Goldschmidt [27] . . .
	industrial practice	Müller [34] . . .	Frankenberger & Badke-Schaub [23], Lloyd & Deasley [32]

Fig. 1. *Emphasis of empirical investigations.*

On the one hand, design processes are investigated by engineers and ergonomic scientists. In these investigations influences of individual prerequisites are not in the focus of interest. On the other hand, individual prerequisites of single subjects are grasped in collaboration with psychologists; their point of interest is especially on individual styles of problem solving and their influences on the design process.

As mentioned above, behaviour strategies of teams solving complex problems are mainly investigated in laboratory situations by psychologists. The investigation of individual- and group-related factors on engineering design practice in cooperation with psychologists represents a new field of research [32, 23].

2. General Aims of the Research Project

Obviously, the effectivity of design processes in industry is, besides the technical problems, determined by several non-technical factors from the fields of the individual prerequisites, the designers' collaboration and the working environment. Thus, we based our investigation on a general starting model of four central influences on the design process in practice: 'Individual prerequisites', 'prerequisites of the group', 'external conditions', and the 'task'. (cf. Fig. 2 with examples).

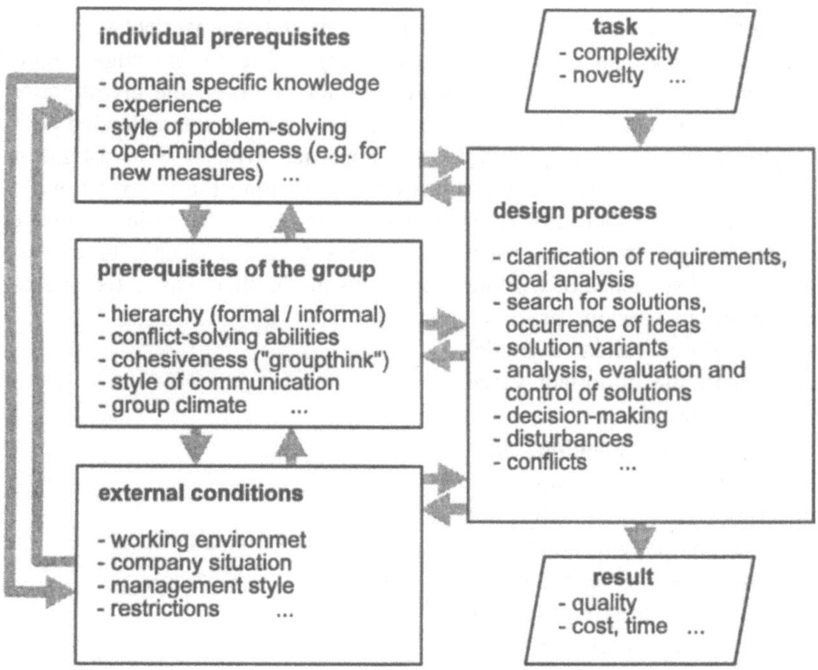

Fig. 2. *Factors influencing the design process and the result.*

In order to support the daily work in engineering design with methodical aids, and in order to improve design education with special emphasis on teamwork, it is necessary to understand the importance and interrelation of the different factors determining design processes.

Therefore, the overall aim of the research project 'Teamwork in Engineering Design Practice' is to identify factors influencing the design process, stemming from the fields of the individual prerequisites, the prerequisites of the group, the external conditions and the task, and then to build a model of cooperative design work in practice based on these findings. This model should describe the interaction of the different factors on the design process and its results and form the basis for further development of systematic design with special emphasis on teamwork.

3. Approach and Procedure

A detailed and exact surveillance of design practice with its vast number of influencing factors makes it necessary to observe the design work without participating: the advantage of this method is that the investigator interferes only minimally with the observed design process and furthermore, has enough time to concentrate on relevant aspects of the design work.

The aim of a differentiated compilation of influencing factors in design practice implies dealing with a large number of dependent and independent variables. The complexity of such an approach makes it impossible to investigate so-called 'comparable groups' in several companies. On the contrary, the investigation has to focus on a very detailed observation of 'single cases' over an extended period of time. Therefore, we chose a single-case approach with two investigations in industry. In the first investigation, the results concerning the interrelations of the different influencing factors were integrated in a model which reflected the central interdependencies of design processes in industry and was able to explain the occurrence and sequence of working steps. This model then was tested in a second investigation in design practice, in order to yield a validated model in which at least the important variables have been reviewed.

Fig. 3. *Pneumatic fruit press.*

The *first* investigation in the research project 'Teamwork in Engineering Design Practice' took place in a company producing agricultural machinery. Over the course of four weeks, the design process of a group of four designers redesigning a fruit press was observed and documented. (cf. Figure 3).

Fig. 4. *Particle board production plant.*

The *second* investigation was conducted in a company of the capital goods industry. In this company we observed three projects of a design team developing and redesigning several components of a particle board production plant (cf. figure 4) for eight weeks.

The high number of factors influencing the design process has to be compiled using a wide variety of investigation methods, in order to guarantee valid results concerning their impacts on the design work. In the following chapters 4, 5 and 6 the most important methods for compiling the 'design process', the 'external conditions', the 'individual prerequisites' and the 'prerequisites of the group' are introduced.

4. Compiling Data of the Design Process and Assessing the External Conditions

The external conditions, such as 'branch' and 'economic situation of the company', its 'culture' and 'organisation', the 'flow of information' and the 'communication' within the organisation, and last but not least, the 'direct working surrounding' with its restrictions, are usually stable during the investigation and can therefore be assessed one at a time.

Contrary to this, the dynamic course of the design process requires a differentiated description of short time spans as determined by the duration of the relevant events. Therefore, the scale of the time intervals was linked to the process-characteristics we use to describe the design work in practice.

We have developed a standardized method for investigating cooperative design work in industry. The basic idea of this method is a parallel and flexible use of both direct and indirect methods to compile data on the design process.

The primary *direct* method is the continuous non-participating observation of the design work. Sitting in the same room, a mechanical engineer observes the activities of the designers in terms of e.g. working-steps, subfunctions/components, ideas and solution variants, etc. A psychologist, in the same room, concentrates his observations on the social aspects of the design process (e.g. ways of decision-making and group interactions). For the detailed documentation of the numerous relevant data describing a team session or important phases of individual work, a standardized laptop-based 'on-line' protocol was used.

This recording via protocol lists provides a description of the design work as a problem-solving process. Among others, we record the category of 'design steps' in accordance with proceeding plans proposed in systematic design [cf. 45, 36]. The development of the technical solution is mainly described by the 'subfunctions' and the 'solution variants' which the designers are working on. Furthermore, we record their collaboration with colleagues and relate every utterance in a discussion to the individual designer who is speaking. Video recordings of all phases of team-work and relevant phases of individual design work enable us to review the description of the design process in specific interesting phases and to render a detailed account of it [cf. 22].

The final protocol lists are transcribed word by word and reflect the entire design process with an average duration of 30 seconds per protocol-line. These protocols are the basis of the following qualitative and quantitative analysis of the design process, using the special software *AnaKonDa*. This program allows, among other features, the analysis of the graphical documentation of each recorded process characteristic in a time-schedule diagram. The time- schedule-diagrams represent the development of the technical solution by changing the subproblems and solution-variants in the course of the design steps. Using these time-schedule diagrams for the analysis of design processes, the activities of the designers can be projected on both a subject level and a process-related level, as figure 5 illustrates.

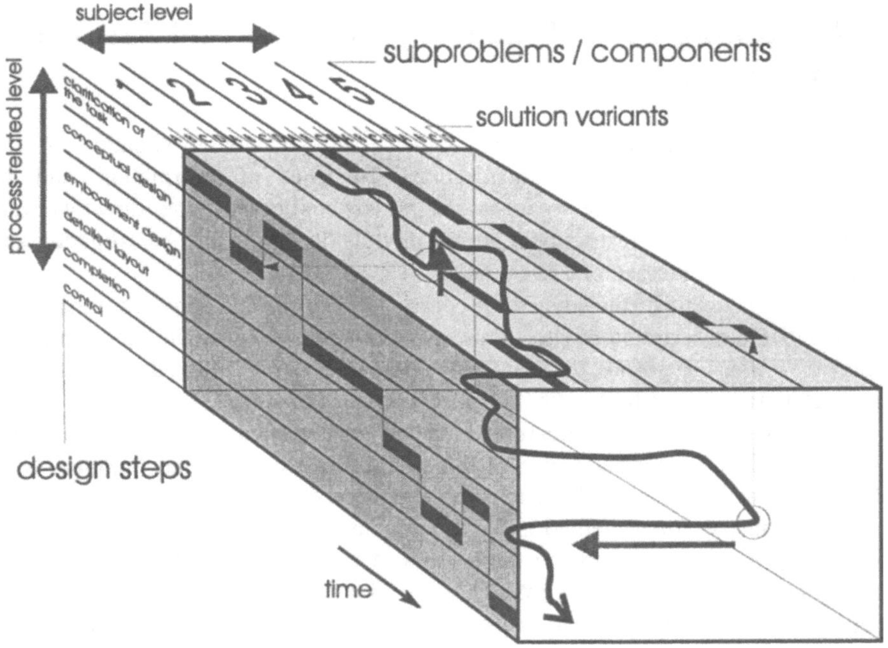

Fig. 5. *Description of a design process through the course of subproblem-solving and solution variants (subject level) and the course of design steps (process-related level).*

In order to fully account for the design process, it is necessary also to evaluate the non-observed work of the engineers involved. Therefore, we use *indirect* methods such as a diary-sheet where the designers can note which subproblem they have worked on, how they have solved problems or when they contacted their colleagues. This sheet can be completed with a minimum of effort in order to avoid a loss of motivation during the investigation.

Moreover, we analyse the documents and ask the designers about their work and their personal opinions on the solution development. These interviews, based on the diary-sheets and the documents, provide important additional findings about the design process and help us to understand the development of the solution and the background of technical decisions. Figure 6 shows the procedure of compiling data of the design process and an excerpt of a revised on-line-protocol.

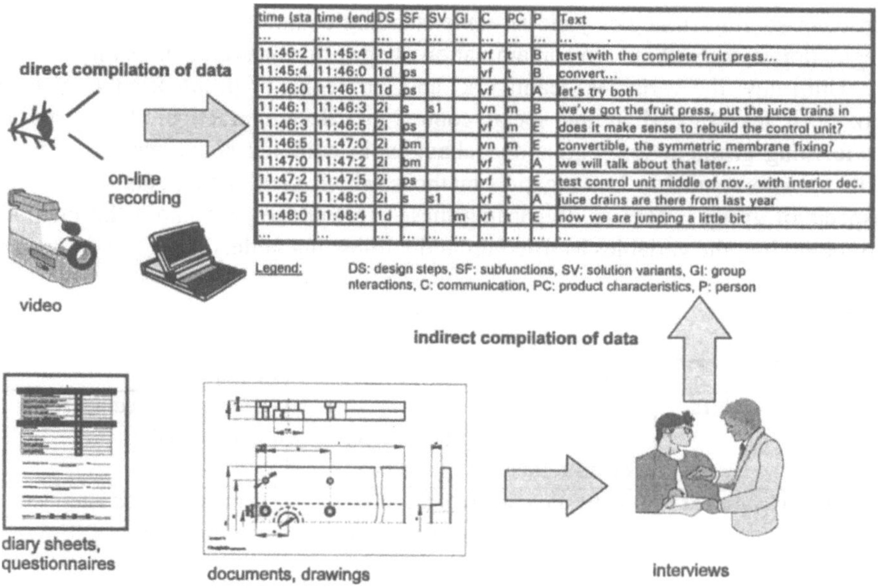

time (sta	time (end	DS	SF	SV	GI	C	PC	P	Text
...	...	...	...	...	...	...	...	...	...
11:45:2	11:45:4	1d	ps			vf	t	B	test with the complete fruit press...
11:45:4	11:46:0	1d	ps			vf	t	B	convert...
11:46:0	11:46:1	1d	ps			vf	t	A	let's try both
11:46:1	11:46:3	2i	s	s1		vn	m	B	we've got the fruit press, put the juice trains in
11:46:3	11:46:5	2i	ps			vf	m	E	does it make sense to rebuild the control unit?
11:46:5	11:47:0	2i	bm			vn	m	E	convertible, the symmetric membrane fixing?
11:47:0	11:47:2	2i	bm			vf	t	A	we will talk about that later...
11:47:2	11:47:5	2i	ps			vf	t	E	test control unit middle of nov., with interior dec.
11:47:5	11:48:0	2i	s	s1		vf	t	A	juice drains are there from last year
11:48:0	11:48:4	1d			m	vf	t	E	now we are jumping a little bit
...	...	...	...	...	...	...	...	...	...

Fig. 6. *Compiling the design process using direct and indirect investigation methods.*

5. Compiling Data on the Individual Prerequisites

It is common knowledge that individual behaviour is influenced by several factors. A reduction of the complex cognitive, emotional and behavioural processes to one or two 'important' characteristics seems almost impossible. People usually behave according to the actual situation, no paradigmes can be considered universally valid for any situation or any person's behaviour, with the exception of a few psycho-psychological theories (e.g. Müller-Lyer-Deception). For example, a person confronted by a novel, complicated problem will take longer to analyse if there is enough time, if the problem is important or if he or she sees a good chance to solve the problem. Obviously, there are many conditions determining the human behaviour in different ways.

The aim of this investigation is to comprehend rules and determining factors of cooperative engineering design work. Therefore, real design processes in industry were taken as a basis for this study. Assuming that design processes are fairly realistic examples for complex problem-solving processes, it is important to look at the behaviour strategies of engineers in complex and novel situations. The investigation of human action regulation and problem solving with computer-simulated complex problems is one main goal of the psychological investigations at the University of Bamberg (see e.g. Dörner [12]; Badke-Schaub & Tisdale, [5]). These computer-simulated problems are characterized by requirements similar to other complex problems, for example like design problems. Such studies on problem-solving strategies yield an understanding of individual and group action-regulation styles without the contamination of specific constraints in reality.

Until now the variables 'teamwork' and 'strategies of problem solving' have not been investigated in the social context of 'design practice'. From studies concerning group-action-regulation behaviour in complex situations [cf. 3, 4, 10, 30], we know suitable investigation methods which yield statements on relevant aspects of group processes. Regarding these prerequisites, we chose the methods for compiling the variables as listed up in the following table:

Table 1. *Variables and methods for compiling the individual prerequisites.*

field of data	variables	methods
biographical data	- age - professional education, career - qualification and experience	- semi-structured interview - questionnaire
work environment	- motivation; job satisfaction - evaluation of the organisation - evaluation of the actual project - relationship to colleagues and to superiors	- semi-structured interview - questionnaire
ability for dealing with complex problems	- analysis and information-gathering - action planning - dealing with time pressure - dealing with stress	computer-simulated microworlds - fire (individual) - machine (individual) - Manutex (group)
special competences	- heuristic competence - social competence	- questionnaire (KF, Stäudel) - observing and analyzing the interactions of the group
abilities concerning the design process	- clarification of the task - search for conceptual solutions - selection and control	- diary sheets / marks-on-paper - on-line protocol of the design process (Video and tapes)

The biographical data and personal evaluations of the working conditions were compiled mainly by means of semi-structured interviews, whereas the ability of dealing with complex problems was compiled by making the person solve different computer-simulated problems.

Contrary to design tasks, these computer-simulated problems can be solved without any specific previous knowledge. These scenarios are used to study the ability of subjects to tackle novel problems in a complex, dynamic and opaque system [cf. 15]. A specific previous knowledge is not intended here, because we are interested in the action-regulation styles of the individuals when being confronted with the specific requirements of different complex situations. Thus, in dealing with these standardized problems, which are isolated from external conditions as far as possible, individual heurisms and strategies become obvious and allow us to make statements on the abilities of a person in dealing with novel, complex situations [5].

Thereby, the behaviour of the subject does not merely yield a single numerical variable (e.g. the 'quality' of problem solving). The planning and proceeding of the subject are rather understood as strategies containing sequences of different variables such as evaluations of questions, decisions, etc. In former investigations it was shown that people show similar strategic behaviour in different complex simulated problems and in design work. These similarities can be interpretated as individual action styles [cf. 1].

In the research project, 'teamwork in engineering design practice', the designers were confronted individually with two novel, complex and dynamic problems ('machine' and 'fire') in order to comprehend the individual thinking and action-regulation in complex situations. A third computer-simulated problem was given to the designers as a group.

These simulations were selected because they require different manners and strategies of action regulation.

1. Machine: The computer simulation 'machine' [41] represents a chemical plant consisting of four different, interconnected production cycles. Each of these cycles produces a special material which is a necessary raw material for the next production cycle. Subjects exposed to this computer simulation have to run the plant trying to obtain the maximum output. Successful running of the plant requires a careful examination of the dependencies and the varying importance of the production cycles before manipulating the parameters such as pumps and valves. Thus, apart from analytical skills, stress resistence is demanded, because the tanks tend to overflow, especially if the subjects are not aware of the interrelations within and between the cycles. This obvious failure can cause emotional reactions resulting in a lack of analysis. All in all, this simulation demands a problem-solving behaviour applicable to design problems. During the design process, too, subfunctions have to be recognized in regard to their connection and importance to the entire function in order to gain the best possible solution.

2. Fire: Another computer-simulated problem called 'fire' was used to examine the designers' ability to adjust their way of action regulation to a changed problem structure [cf. 13, 20]. Therefore, the problem 'fire' provokes almost contrary problem-solving patterns of behaviour than the 'machine' simulation.

In this computer simulation the subject takes the role of the chief of a fire brigade in the Swedish forests. His task is to fight fires by coordinating the fire units (fire engines and helicopters). Again, this simulation represents a dynamic, complex problem with multiple goals (several fires can light up simultaneously, they can endanger villages and wooden dams at the same time, etc.). The scenario was given three times in different variants: the first two games were very similar concerning their difficulties, whereas the third game included extreme requirements, being very difficult due to the many simultaneously lighting fires and thus giving rise to a stress situation. Nevertheless, in each game it is essential to notice and to consider the important factors influencing the spread of the fire, such as velocity and direction of the wind. Moreover, it is important to realize quickly where to place the units in an optimal way, which fire requires the helicopter, which units have to be filled up with water, whether a dam or a village is endangered by the fire, etc. Thus, the computer simulation 'fire' allows the researcher to compile the subject's abilities 'flexibility' and 'planning of long-term and side effects' under the condition of time pressure.

Over and above the analysis of abilities in the defined computer-simulated situations, the specific competences of the designers were compiled during the design work. On the one hand, the activities of each designer, their difficulties, the ways of solving their problems and the influence of colleagues were noted in diary-sheets. On the other hand, the design work was recorded by on-line protocols, and important phases of the design process were recorded on video and afterwards video sequences were analysed in detail and categorized (see Chapter 4).

Another important view on the designers' work was their self-assessment concerning the own problem-solving abilities (e.g., heuristic competence), assessed by a questionnaire developed by Stäudel [42]). Several studies indicate that a positive self-assessment of problem-solving abilities by the individual supports successful problem-solving in complex situations [cf. 42], whereas a significant deviation from the average (to both sides) allows interesting conclusions on the self-assessment of a person [2].

Additionally, the social competence of the individual designer was evaluated, based on the observations of group activities, both in design practice and in the simulations. The aim was to comprehend the individuals' abilities to guide the group or to become integrated into the group. Of course, this integration can vary extremely, according to the various characteristics of the group and of the individual designer, such as role, state and communication.

6. Compiling Data on the Prerequisites of the Group

Similar to the compiling of the specific individual prerequisites and the analysis of the subjects' behaviour, the different characteristics of the groups were compiled. On the one hand, we intended to look at the structuring aspects of the group such as role-taking behaviour and leader functions; on the other hand, group processes were observed and analysed on the basis of specific event protocols.

Consequently, we chose group interaction processes during the design processes and described them in terms of individual and group behaviour patterns.

Another important diagnostic situation was the computer-simulated scenario Manutex [40], which had to be performed by the designer group. Manutex is a small garment factory in Malaysia, that has been experiencing severe problems since the death of its founder. The task of the engineer-group is to operate this factory, which is simulated in detail on the computer. The group is free to allocate the total time of approximately four hours to run the company for 24 simulated months. The team of engineers has to act as a top management group and has to make decisions with respect to manufacturing, marketing, materials, financial, personnel and administrative areas.

This situation combines the requirements of acting in a complex situation and organizing a group in a complex setting abroad from the 'normal' designers work. Whereas the problem-solving activities demand a high extent of goal-analysis and emphasizing priorities, the group situation causes the necessity for each individual to express own ideas and strategies of proceeding. Getting his or her own suggestions accepted is linked to different characteristics of the individual, mainly the concept of social competence, which includes several abilities of acting in groups (e.g. the ability to cooperate, the ability to communicate, etc)

As described above, the analysis of the computer problem-solving situation helps us to examine how the individual and the group dealt with the various requirements concerning complex problem solving and group organization. The questions we are interested in cover two main areas: the first question concerns the structure and the organization of the group during the problem solving process; the second question is how the group approaches the problem in terms of behavioural patterns which may be reponsible for producing the observed results.

The simulation situation provided us with several important individual and group specific characteristics. These data were compared with the analysis of different observation-periods of the design process. Both situations were categorized according to the same protocol-system, which was developed on the basis of the phases of the action regulation, according to Dörner [12]. Additionally, socio-emotional acts and organizational aspects concerning the group were protocolled.

On the basis of the categorization system, each individual speech act was protocolled so that the actual as well as the ongoing process of the action-regulation of the individual and the group was registered. The evaluation of the different categories gave indication about the individual's role concerning behaviour patterns and group processes. For example, if Engineer 'A' is the one who is responsible for the information-gathering in the group, then we will get a constant high number of specific questions; we will find another pattern of questions if Engineer 'B' cares about the group organization. Changes in the allocation of the different tasks was observed in the evaluation and development of the different categories over time. Additionally, we have a protocol of the individual behaviour and communication pattern as well as of the behaviour pattern of the group.

The compilation of the characteristics of the individual and the group is shown in table 2. The table demonstrates the differentiation between structure and process characteristics on the one hand, and the group and the elements of the group on the other hand.

Table 2. *Classification of structural and process characteristics of individual and group [35]*

	structure	process
	characteristics of the group	**events (design process, simulation)**
system	for example: goals, leadership, social roles, communication patterns, task allocation	for example: action regulation: information-gathering, analysing the problem, planning, decision making, self organization - group organization - concerning simulated problems - concerning the design problems
	characteristics of the group members	**individual behaviour patterns (design process and simulation)**
elements	for example: biographical data (education, career, etc.), knowledge, experience, needs, competence, abilities	for example: action-regulation styles concerning complex problems, action-regulation styles under stress conditions, action-regulation styles concerning 'critical situations'

All in all, the data of the different domains - design process and result, external conditions, prerequisites of the individual and prerequisites of the group - are recorded on the basis of both the self-estimation of the engineers and the estimation of the observers, as shown in Table 3.

Table 3. *Methods for compiling data in the domains of influencing factors.*

methods	**domains of influencing factors**			
	design process and result	external conditions	prerequisites of the individual	prerequisites of the group
interviews	•	•	•	•
on-line-protocols	•	•	•	•
diary sheets	•	•	•	•
marks-on-paper	•		•	
questionnaires			•	•
computer-simulated problems			•	•

7. Evaluation and Modelling

7.1 Distinguishing between 'critical situations' and routine work

The preparation and evaluation of the extensive data in the first investigation called for a new, standardized procedure. It was necessary to find an approach for connecting the data from the fields of the design process, the external conditions, the individual and the group that would allow for both the proof of the connections and the generalization of the findings.

We developed a method to connect the detailed data of the design process and the influencing factors in a way that allows to check the connections. As mentioned above, the design processes are documented in much detail. If we take a closer look at the design work, we see the elaboration of goals, the search for and generation of solutions, analysis, decision-making and control processes, nearly at all times and on each level of concrete realization. The investigations so far confirm that not every little decision in detail design is important for the solution development and therefore signals 'routine work' for the designer. Hence, if we take a more abstract view of the design process, we can identify phases of routine work on the one hand and 'critical situations' on the other hand, where the design process takes a new direction on a conceptual or embodiment design level. These 'critical situations' are thus determining the following course of work and its result.

The basic idea of the evaluation of the design process by abstraction and reduction on phases which are interesting for the further analysis is illustrated in figure 7.

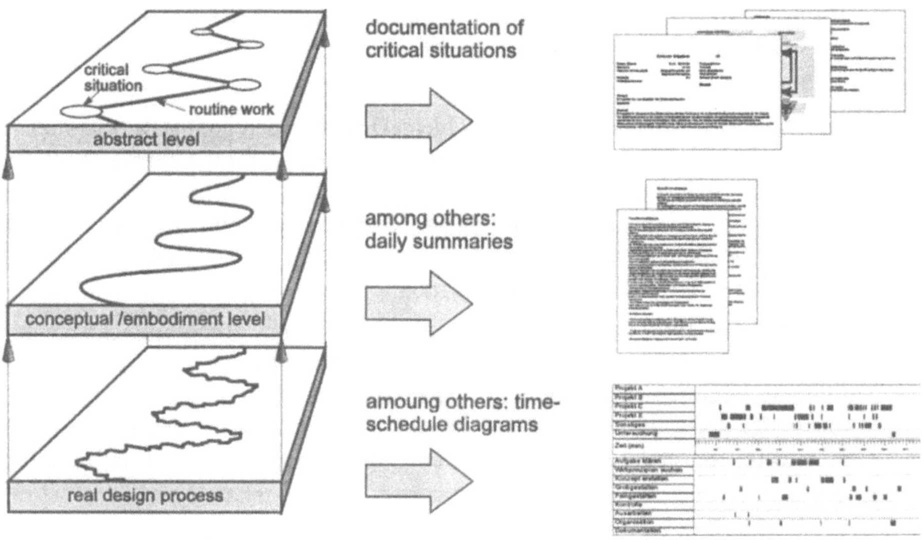

Fig. 7. *Standardized evaluation of the design process on an abstract level according to 'critical situations' (knots) and routine work (lines).*

The evaluation follows defined steps. Based on the detailed protocol-lists and their graphic representation in time-schedule-diagrams, the written summaries of the daily work in 'day-sheets' mean a first reduction on the essential course of work. Then the critical situations[2] are identified in the protocols and day-sheets and classified according to defined rules. These rules are deducted from the general problem-solving cycle [cf. 19] and the requirements of solving complex problems [cf. 12] (see figure 8).

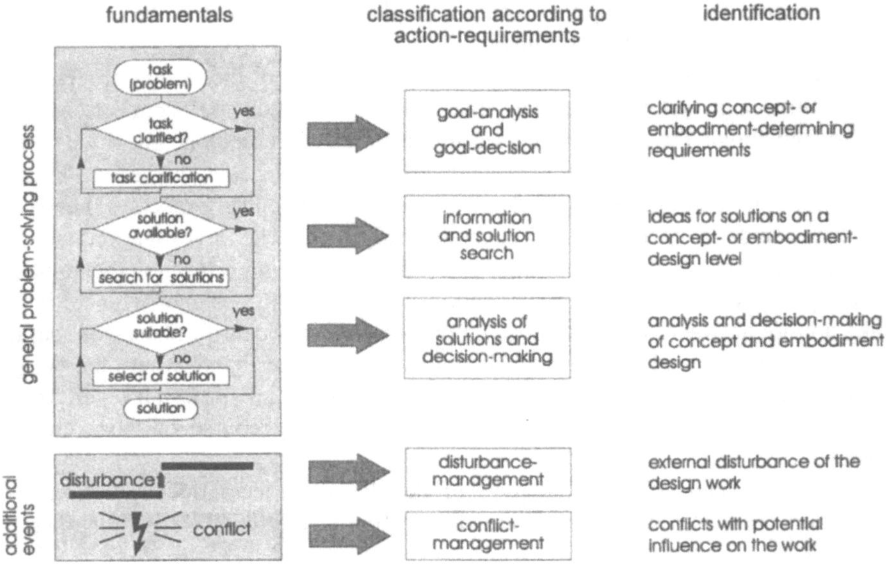

Fig. 8. *Classification of 'critical situations' according to the general problem-solving process [cf. 19], and the action-requirements [cf. 12].*

1. **Goal analysis:** Mostly, the given problems have an open goal state, that means the situation is not able to give any concrete instruction for the course of action. Analysing goals as well as defining precise criteria are necessary for a first determination of the further problem-solving process. Whether the mode of behaviour in coping with goal analysis is successful or insufficient, each goal elaboration is defined as a critical situation.
2. **Information and solution search:** Generally both processes, search for information and solution search are linked together, therefore we put the two different requirements into one category. Confronted with a problem, knowledge about the structure and the variables of the system is necessary. On a concept- or embodiment-design level the beginning of an information

[2] This method of 'critical situations' sounds familiar by reference to the 'critical incidents' by Flanagan (1954) or the 'critical moves' by Goldschmidt (1996), but it follows another concept, because the identification of the critical situations takes place according to the requirements of the design process.

gathering process or an idea-generation process are defined as a critical situation.

3. **Analysis of solutions and decision making:** In the same way, analysing a solution variant and making a decision are closely connected during the control of a solution. This process of analysing and decision making is an important part of the design process, because it determines the further design process and the result in a positive or negative way.
4. **Disturbance-management:** Disturbances from outside can cause changes in the course of design work, such as changing the actual subproblem without having finished the work on it. Disturbances require a suitable 'disturbance-management' by the designer; they may or may not be managed appropriately, in any case they are defined as critical situations.
5. **Conflict-management:** Conflicts are defined as problems in the interpersonal area. Conflicts may occur in the own team or with people of external companies. Such conflicts may have an effect on the individual course of the design work or on the working team. Depending on the quality of 'conflict-management', conflicts may have a more or less important impact on the process, in any case conflicts are defined as critical situations.

7.2 Establishing the model

'Critical situations' are defined as situations where the design process is determined decisively. Therefore, such situations are of special interest in isolating the most essential influences on the design process. In order to extract these influences and to explain the effect of a critical situation, we have built a submodel of interrelations between the influencing factors and the process characteristics for each of these relatively short 'critical situations' (see figure 9). Each identified relation was substantiated separately. Specific interviews with the designers, combined with video-feedback of selected 'critical situations', helped to revise the submodels.

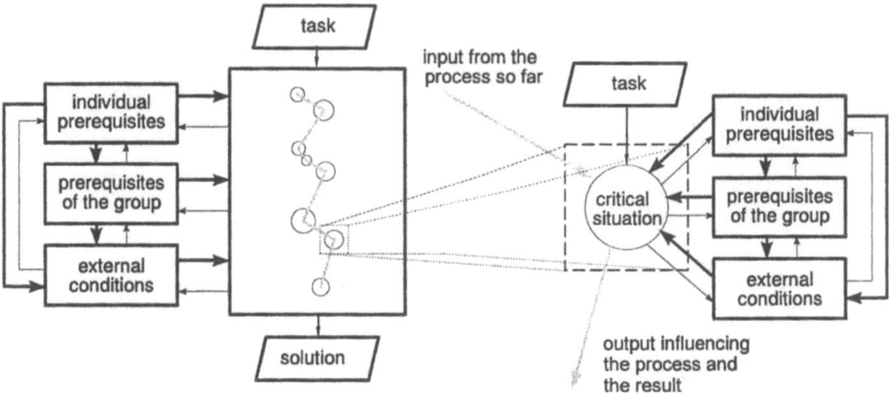

Fig. 9. *Influences on the design process as influences in 'critical situations'.*

The sum of the different interrelations in the individual submodels, led us to a model of relations between influencing factors and process characteristics in all 'critical situations' of the design process. We identified altogether 265 critical situations in the four analysed projects of the two investigations and explained the course of work by more than 2200 single interrelations between factors, process characteristics and the result. The reduction to at least 34 different influencing factors illustrates the suitability of the model. This model is based on a high number of single cases (critical situations), in which every statement is explained by data and can be checked.

Figure 10 illustrates the combination of the submodels of each critical situation to an entire model of interrelations (model of the first investigation).

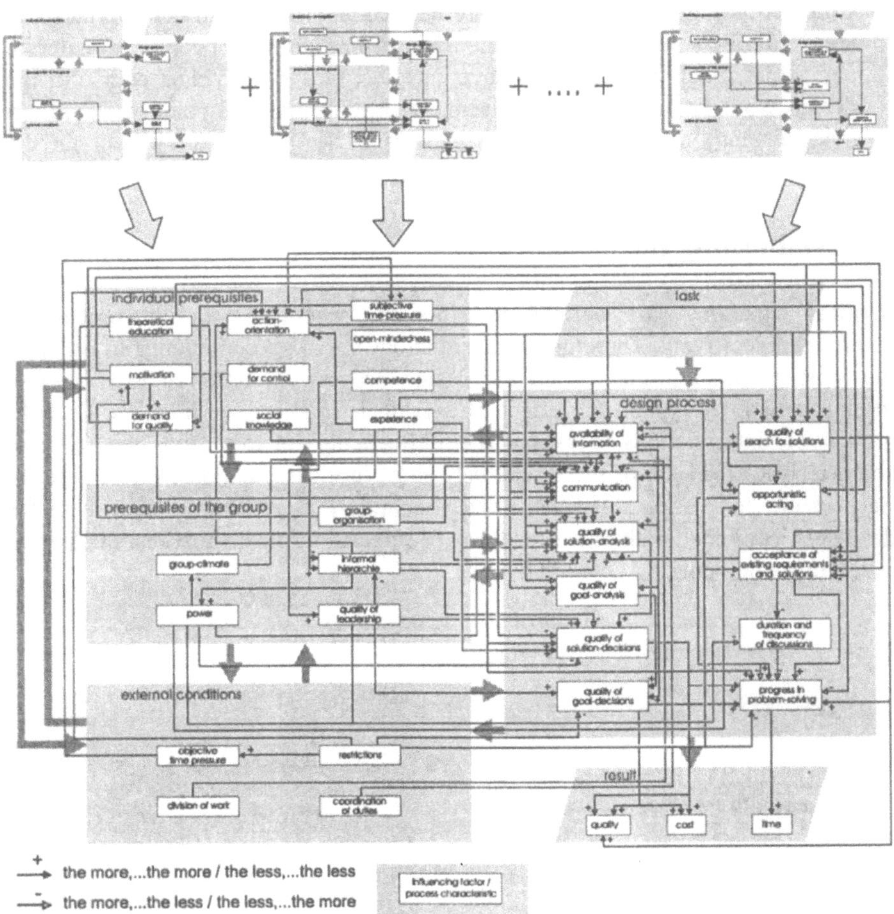

Fig. 10. *Model of relations in the entire design process based on 62 'critical situations' in the first investigation.*

7.3 Analysing the model

In the model of the relations of the entire design process as shown in figure 10, each relation is represented only once. In the first step, the importance of the influencing factors and their interrelations can be evaluated by their frequency of occurrance in all 'critical situations' in the four design projects. But as figure 8 shows, each type of critical situation has a specific role in the design process. Therefore, in order to make more specific statements on the central mechanisms leading to success or deficiency in the design processes, we have to sum up the models of each type of critical situation seperately. Figure 11 shows how often the different types occurred with positive and negative consequences:

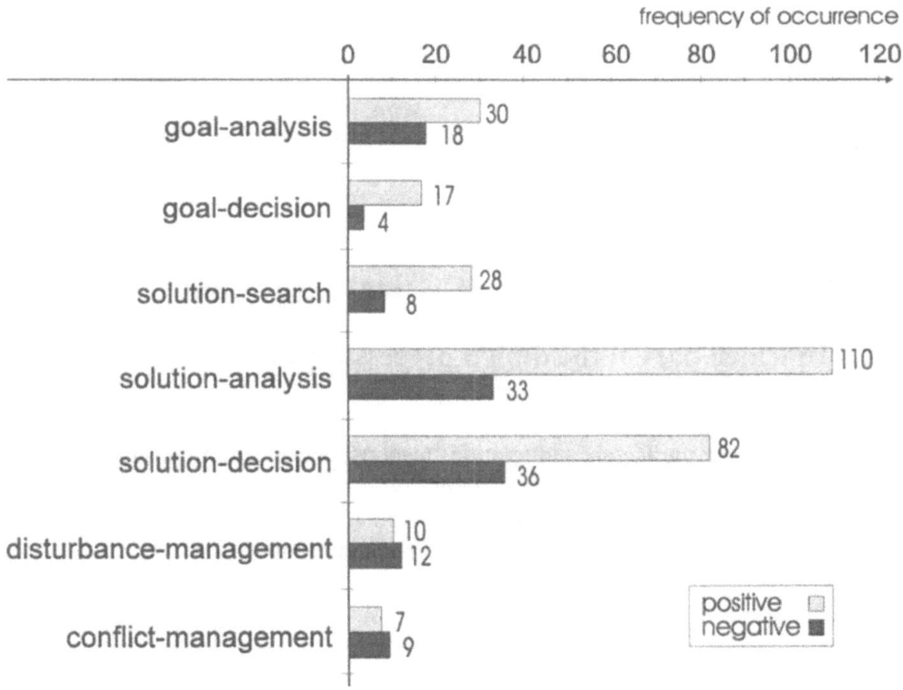

Fig. 11. *Frequency of occurrence of the different types of critical situations summarizing the four projects (the sum is more than 265 because analysis and decision often occur together in one situation).*

By summing up the single models of each type of critical situation, we can identify the central mechanisms responsible for positive or negative outcomes of the different types of situations. On the basis of this analysis we can answer questions such as, 'which are the main factors responsible for a deficient analysis of goals?' or 'which are the mechanisms leading to low quality'?

Before presenting results, we want to discuss if it is possible to quantify generally accepted determinants and relations on the basis of the data of two companies and four projects. Figure 12 depicts the number of the influencing factors which were necessarily added in order to explain the different critical situations in the special project. The percentage of newly added influencing factors (as well as the percentage of the relations) decreases from project to project. Furthermore, the most important relations occurred in all four projects very often. This result leads us to the assumption that within the four projects we captured the most important influencing factors and relations.

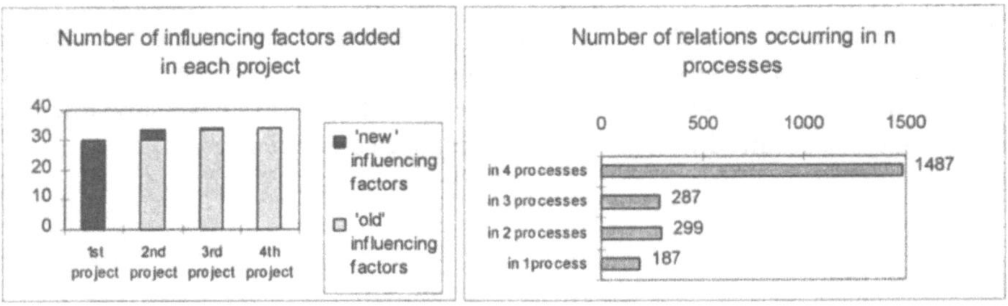

Fig. 12. *Number of infuencing factors added in the four projects and number of relations occurring in n processes.*

8. Design Productivity in 'Critical Situations'

8.1 In which part of the design process is productivity determined?

In order to detect the factors and relations responsible for 'design productivity', we need to know how important the different types of critical situations are for the result in terms of quality, time and cost. Figure 13 shows for each type how often the situations influenced the result in a positive and negative way.

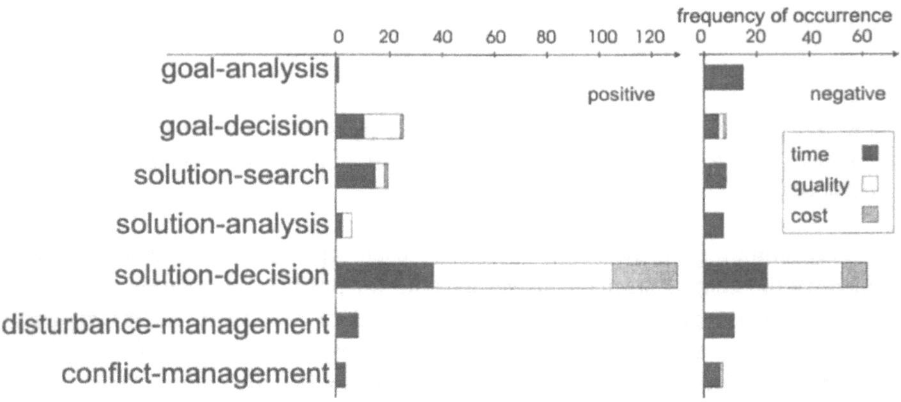

Fig. 13. *Importance of the critical situations for the result.*

Obviously, especially the decisions of solutions are important for the result. 'Quality' and 'cost' of the result are mainly determined by decisions of solutions, whereas the more process related characteristic 'time' is influenced in each type of critical situation.

What are the main factors and relations leading to good or poor quality, cost and time? In order to answer these questions we analyse the critical situations with relation to the result. As an example, the central mechanisms leading to low quality, high cost, and savings in time are presented in the following.

8.2 Mechanisms leading to low quality

Problems with quality were mainly determined by deficient decisions; only in a few situations of goal decisions could we detect a direct influence on the quality of the result. Consequently, the factors and relations as shown in figure 13 occur mostly in situations where designers are deciding which solution they want to further develop. Figure 14 describes the main mechanisms responsible for *low quality* in the four projects. The thickness of the **arrows** depicts the frequency (in percent) of the relations occurring in all situations with bad influence on quality. The thickness of the **frames** depicts the frequency (in percent) of the factors identified in these critical situations.

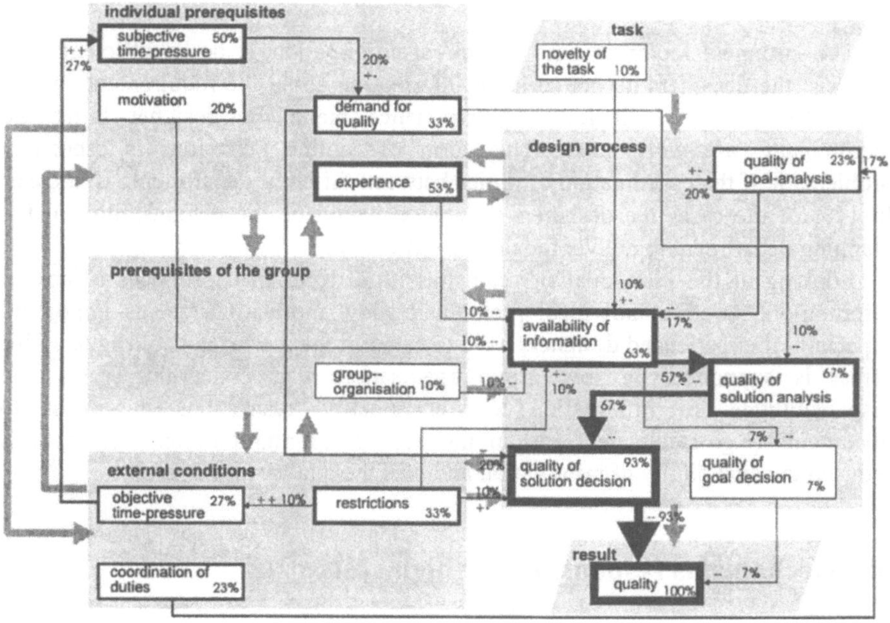

Fig 14. *Factors and relations leading to low quality of the solution (in 30 situations).*

Figure 14 shows that the quality of the solution decision mainly depends on the quality of the former analysis. Nevertheless, certain external conditions also influence the quality of solutions in a negative way: Restrictions (e.g. problems with suppliers) can cause objective and subjective time pressure, which decreases the demand for quality of the designer. Thus, the solution analysis is reduced and suboptimal solutions are chosen. However, the most importanat variable is the *availability of information.* A lack of information concerning the demands and the solution leads inevitably to a deficient solution analysis. Figure 15 shows the factors reducing the availability of information.

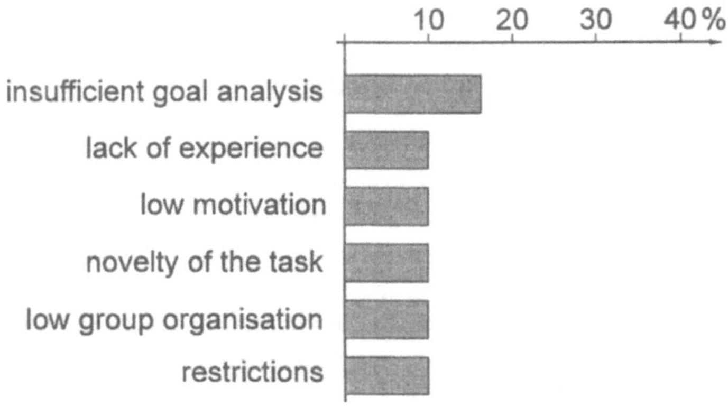

Fig. 15. *Factors reducing the availability of information.*

The strongest factor causing a lack of information is an insufficient goal analysis: the designers do not seek for information about the requirements of the task. Mostly their high level of experience make the designer think the requirements are well-known. This form of routine behaviour is especially problematic if the coordination of duties between different departments is unclear: In 17% of the cases the designers thought it would be the responsibility of the planning department to deliver the task specification.

Looking at the individual prerequisites little experience produces a lack of information. Another individual prerequisite, low motivation, seems important, especially if experienced designers have to support inexperienced colleagues. This effect is supported by low group-organisation, e.g. concerning rules and responsibilities for information transfer. Finally, restrictions in access to documents (e.g. catalogues) and in the use of information-management tools reduced the availability of information and thus the quality of the result.

8.3 Mechanisms responsible for high costs

In the four projects altogether 13 situations with direct negative consequences for the cost of the product were identified. Again, these situations were mainly decisions of solutions, but we also found a situation of conflict-management and goal decision. In spite of the fact that these are relatively few situations with direct decisions for a more expensive solution, costs are also indirectly influenced by 76 situations leading to time-delays (see chapter 8.4). Figure 15 shows the most important factors and relations leading directly to higher costs:

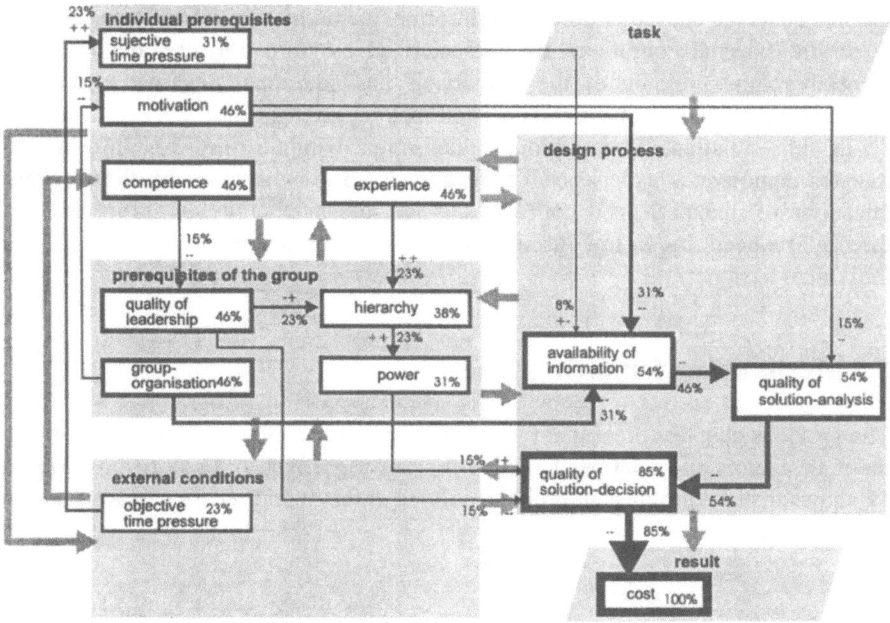

Fig. 15. *Factors and relations leading to higher costs of the solution (in 13 situations).*

Deficient decisions of solutions leading to high costs are again mainly caused by an insufficient analysis, which is very often the result of a lack of information. Besides this very common mechanism, the group prerequisites, and especially the managerial aspects, have a crucial influence on decisions. Figure 16 shows the main factors leading to deficient decisions with negative effects on costs.

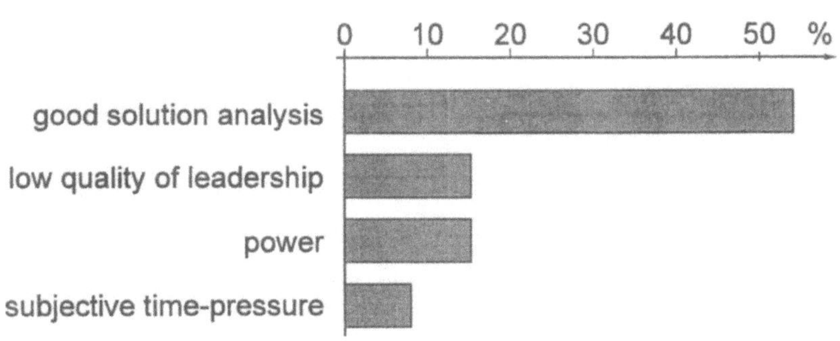

Fig. 16. *Factors leading to deficient decisions of solutions concerning costs.*

The quality of leadership is often a question of competence. Low quality of leadership has two effects: On the one hand, low competence of the leader can lead directly to wrong decisions. On the other hand, it increases the power of an informal hierarchy. This power also leads to decisions which are not based on rational analysis, but rather power and informal hierarchy suppress free discussion.

If the external conditions are characterized by time pressure, the resulting subjective time pressure of the individuals has additional negative effects on information transfer, analysis and decision-making activities.

In this situation, a low group-organisation without rules for information transfer demotivates experienced colleagues, who are disturbed by the frequent questions of unexperienced colleagues. Thus, the availability of information is directly reduced by a low group-organisation and a low motivation of the individual designer.

8.4 Mechanisms for saving time

The previous chapters presented two negative examples, now we want to introduce the main factors and relations having a time-saving effect in 72 situations. Figure 17 shows that the progress in problem solving is the main process characteristic to reach the time goal.

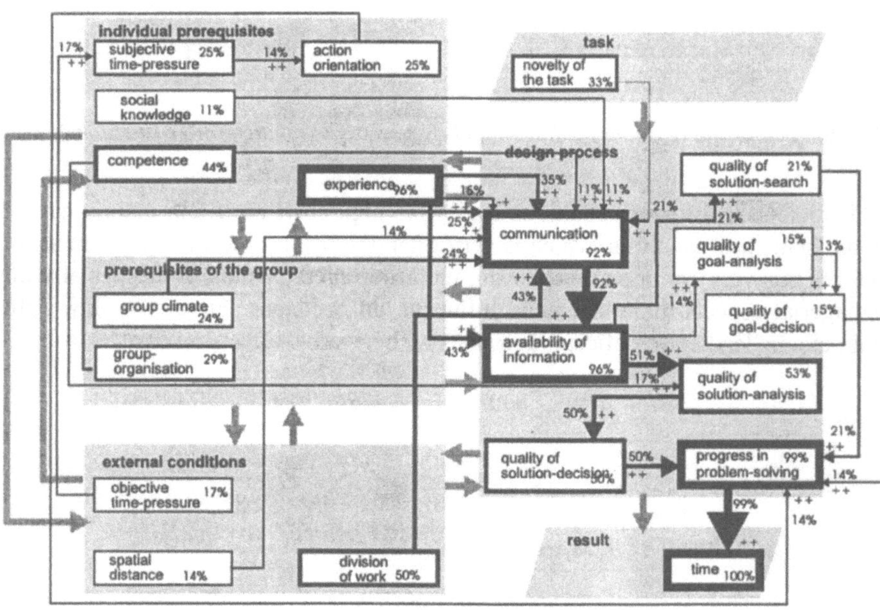

Fig. 17. *Factors and relations with a time-saving effect in 72 situations.*

This progress is triggered by good decisions in regard to goals and solutions and by an effective search for solutions. Additionally, in terms of individual prerequisites, a high action orientation of the designer increased by time pressure pushes the process forward. Figure 18 shows the main factors directly influencing the progress in problem solving.

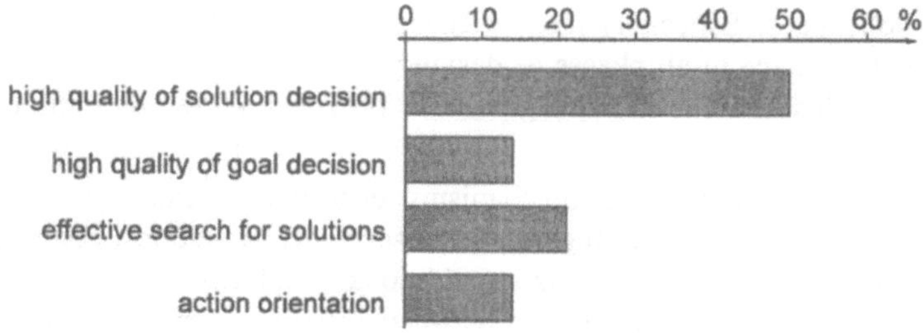

Fig. 18. *Factors influencing the progress in problem solving.*

The main factor for a good decision is a sufficient analysis, which is mainly based on the availability of information. An additional individual prerequisite for a good analysis is the competence of the designer to deal with this information. Again, the availability of information is the main variable in the process. In 45% of these situations information is based on the experience of the designer, but in 85% information is provided by communication with colleagues. Therefore we need to know which factors lead to a successful information transfer. Figure 19 shows the factors leading to communication:

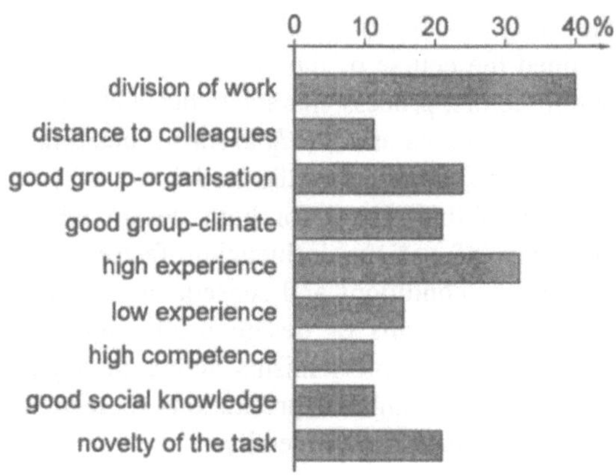

Fig. 19. *Factors leading to communication.*

Communication in modern product development is mainly caused by the division of work. If there are different departments involved in the design process, they need to talk together, they need to exchange information. Thus, the spatial distance to the colleagues is another communication-supporting external condition.

In regard to the prerequisites of the group, a good group-organisation leads to communication, for example, in order to inform the leader about new ideas or decisions. A good group climate itself can be the reason why designers talk more about their design problems than absolutely necessary.

The main individual prerequisite causing communication is a gap of experience between colleagues: Especially inexperienced designers have to ask experienced colleagues for advice in all phases of design work. Thereby, the designer's social knowledge helps to find a colleague with previous knowledge in a specific problem.

The presented time-saving mechanisms occurred in each type of critical situation, as figure 13 shows. Regarding the successful situations, communication was the key factor. This result puts an additional emphasis on the importance of the factors supporting communication between colleagues, and therefore a special emphasis on group-related factors. In spite of the fact that the designers worked for an average of 80% of their time individually, the importance of the group-related factors in 'critical situations' becomes clear by the fact that 88% of the 'critical situations' took place in collaborative work of the designers. Obviously, communication especially takes place in 'critical situations' whereas routine work predominates in phases of individual work.

9. Discussion

In two companies we observed four design projects for altogether 28 weeks. Based on the data of the observed design work we identified altogether 265 critical situations and explained the course of the design work by more than 2200 single interrelations between factors, process characteristics and the result. The reduction to at least 214 *different* relations and 34 *different* influencing factors and process characteristics illustrates the suitability of the model.

What is the benefit of this model for both design education and practice? Knowing the interaction of various influencing factors in different situations provides an insight into the conditions and consequences of relevant factors in the design process, not only in terms of design productivity. This knowledge on important positive and negative mechanisms acting in a particular situation of design work helps to develop suitable precautions in a company and allows a practical relevant design education at university.

References

1. Auer P & Frankenberger E 1994 Vorgehensstile beim Konstruieren - Flexibilität und Invarianz beim Lösen unterschiedlicher Probleme. University of Bamberg, Lehrstuhl Psychologie II, *Memorandum Nr.11*
2. Badke P 1986 Persönlichkeit und Problemlösen: Charakteristika des Verhaltens in komplexen Problemsituationen und Möglichkeiten der Prognose aufgrund von persönlichkeitsspezifischen Faktoren (*unpublished diploma thesis*), University of Bamberg

3. Badke-Schaub P 1993 *Gruppen und komplexe Probleme: Strategien von Kleingruppen bei der Bearbeitung einer simulierten AIDS-Ausbreitung.* Peter Lang, Frankfurt a.M
4. Badke-Schaub P & Buerschaper C 1994 Das Schicksal der Manutex. Gruppenproblemlösen in Ost und West. In: Strohschneider S (ed) *Denken in Deutschland.* Huber, Bern
5. Badke-Schaub P & Tisdale T 1995 Die Erforschung menschlichen Handelns in komplexen Situationen. In: Strauß B & Kleinmann M (eds) *Computersimulierte Szenarien in der Personalarbeit.* Verlag für angewandte Psychologie, Göttingen
6. Beitz W 1983 *Neue Arbeitstechniken beim Konstruieren.* VDI-Berichte Nr. 492. Düsseldorf, VDI-Verlag
7. Beitz W & Luczak H 1989 Vergleichende Belastungs- und Beanspruchungsanalyse konventioneller und rechnergestützter Konstruktionsarbeit. *Forschungsbericht 1987-1988: Forschergruppe Konstruktionshandeln, Nichttechnische Komponenten des Konstruktionshandelns bei zunehmendem CAD-Einsatz*
8. Birkhofer H 1991 Methodik in der Konstruktionspraxis - Erfolge, Genzen und Perspektiven. In: Hubka V (ed) 1991 *Proceedings of ICED 91.* Edition Heurista, Schriftenreihe WDK, Zürich
9. Blessing L T M 1994 *A Process-Based Approach to Computer-Supported Engineering Design.* Thesis, University of Twente, Enschede, the Netherlands. Cambridge, Black Bear Press
10. Brauner E 1994 *Soziale Interaktion und mentale Modelle. Planungs- und Entscheidungsprozesse in Planspielgruppen.* Waxmann, Münster
11. Christiaans H, Dorst D & Cross N 1993 Levels of Competence in Product Designing. In: Roozenburg N (ed) 1993 *Proceedings of ICED 93.* Schriftenreihe WDK. Zürich, Edition Heurista
12. Dörner D 1996 *The Logic of Failure.* Metropolitan Books, New York
13. Dörner D & Pfeiffer E 1991 Strategisches Denken, strategische Fehler, Streß und Intelligenz. *Sprache und Kognition, 11,* 75-90
14. Dörner D, Ehrlenspiel K, Eisentraut R & Günther J 1995 Empirical Investigation of Representations in Conceptual and Embodiment Design. In: Hubka V (ed) 1995 *Prodeedings of ICED 95.* Schriftenreihe WDK 23. Zürich, Edition Heurista
15. Dörner D & Wearing A J 1995 Complex problem solving: Toward a (computer simulated) theory. In *Complex Problem Solving: The European Perspective* by P.A. Frensch & J. Funke (Eds.), Lawrence Erlbaum Associates, Hillsdale, NJ
16. Dylla N 1991 *Denk- und Handlungsabläufe beim Konstruieren.* Hanser, München
17. Ehrlenspiel K & Dylla N 1991 Untersuchung des individuellen Vorgehens beim Konstruieren. *Konstruktion, 43, 43-51*
18. Ehrlenspiel K 1993 Industrieprobleme und nötiges Wissen bzw. Können im Bereich Entwicklung und Konstruktion. *Konstruktion* vol. 45, pp. 389-396
19. Ehrlenspiel K 1995 *Integrierte Produktentwicklung. Methoden für Prozeßorganisation, Produkterstellung und Konstruktion.* Hanser, München

20. Eisentraut R & Günther J 1996 Individual styles of problem solving and their relation to representations in the design process. Proceedings of the First International Symposium on Descriptive Models of Design, Istanbul, 1-3 July, 1996
21. Flanagan J C 1954 The Critical Incident Technique. *Psychological Bulletin, 51, 4*, 327-358
22. Frankenberger E & Auer P 1996 Standardized Observation of Teamwork in Design. *Research in Engineering Design*. Springer, London
23. Frankenberger E & Badke-Schaub P 1996 *Modelling Design Processes in Industry - Empirical investigations of design work in practice*. Proceedings of the First International Symposium on Descriptive Models of Design, Istanbul, 1-3 July, 1996
24. Fricke G & Pahl G 1991 Zusammenhang zwischen personenbedingtem Vorgehen und Lösungsgüte. In: Hubka V (ed.), *Proceedings of ICED 91*. Schriftenreihe WDK. Zürich: Edition Heurista
25. Fricke G 1993 *Konstruieren als flexibler Problemlöseprozeß - Empirische Untersuchung über erfolgreiche Strategien und methodische Vorgehensweisen beim Konstruieren*. VDI-Verlag, Düsseldorf
26. Frieling E & Derisavi-Fard F 1990. Ändert die CAD-Technik die Arbeitstätigkeit von Konstrukteuren? *Zeitschrift für Arbeits- und Organisationspsychologie, 34/3, 135-148*
27. Goldschmidt G 1996 The Designer as a Team of One. In Cross N, Christiaans H & Dorst K (eds.), *Analysing Design Activity*. New York: John Wiley & Sons
28. Haberfellner R, Nagel P, Becker M, Büchel A & Massow H von 1992 *Systems Engineering: Methodik und Praxis*. Verlag Industrielle Organisation, Zürich
29. Hales C 1987 *Analysis of the Engineering Process in an Industrial Context*. Cambridge: University, Dissertation
30. Janis I L 1972 *Victims of Groupthink*. Houghton Mifflin, Boston
31. Janis I L 1982 Counteracting the Adverse Effects of Concurrence-seeking in Policy-Planning Groups: Theory and Research Perspectives. In Brandstätter H, Davis J H & Stocker-Kreichgauer G (eds.), *Group Decision Making*. Academic Press, London
32. Lloyd P & Deasley P 1996 *Ethnographic description of design networks*. Proceedings of the First International Symposium on Descriptive Models of Design, Istanbul, 1-3 July 1996
33. Minneman S & Leifer L 1993 Group Engineering Design Practice: The Social Construction of a Technical Reality. In: Roozenburg N (ed) 1993 *Proceedings of ICED 93*. Schriftenreihe WDK. Zürich: Edition Heurista
34. Müller J 1989 Möglichkeiten und Ergebnisse der analytischen Darstellung konstruktiver Entwurfsprozesse im aktivitäts- und ereignisorientierten Graph. *Konstruktion, 41, 25-34*
35. O'Connor G 1980 Small Groups. A General System Model. *Small Group Behavior, 11*, 145-174
36. Pahl G & Beitz W 1996 *Engineering Design*. Springer, London

37. Radcliffe D & Slattery P 1993 Video as a Change Agent in a Cross-Discipline Design team. In: Roozenburg N (ed) 1993 *Proceedings of ICED 93*. Schriftenreihe WDK. Zürich: Edition Heurista
38. Rothe H-J 1990 *Erfassung und Modellierung von Fachwissen als Grundlage für den Aufbau von Expertensystemen.* Kassel: Gesamthochschule, Habilitationsschrift
39. Schaub H 1988 Manutex: Instruktion, Versuchsleiterhinweise und Hilfstexte. Bamberg: *Interner Bericht am Lehrstuhl Psychologie II der Universität Bamberg*
40. Schaub H 1990 Maschine: Instruktion, Versuchsleiterhinweise und Hilfstexte. Bamberg: *Interner Bericht am Lehrstuhl Psychologie II der Universität Bamberg*
41. Stäudel T 1987 *Problemlösen, Emotionen und Kompetenz. Die Überprüfung eines integrativen Konstrukts.* Roderer, Regensburg
42. Stauffer L & Ullman D 1991 Fundamental Processes of Mechanical Designers Based on Empirical Data. *Journal of Engineering, 2, 1, 91-101*
43. Upmeyer A 1988 Computerunterstützte Diagnosesysteme für die Bewertung von CAD-Software. *Forschungsbericht 1987-1988: Forschergruppe Konstruktionshandeln, Nichttechnische Komponenten des Konstruktionshandelns bei zunehmendem CAD-Einsatz*
44. VDI-Richtlinie 2221 1993 *Methodik zum Entwickeln und Konstruieren technischer Systeme und Produkte.* VDI-Verlag, Düsseldorf

Computer Supported Co-operative Product Development Using a Process-Based Approach

Edward H. McMahon, Ph.D., PE
College of Engineering and Computer Science,
The University of Tennessee at Chattanooga
615 McCallie Avenue Chattanooga, TN 37403 USA

Abstract

The paper describes a computer supported product development system using a process-based approach. The process-based model was implemented on a small cluster of microcomputers using a multi-user database and tested in a graduate class on Product Development. The results suggest that the process-based model can be successfully applied to a wide range of functional activities and different levels of detail.

Several benefits were identified by the users including simultaneous processing of information, equality of input, documentation of decision rationale, a common structure for various functional areas, and the availability of information for review. Several issues were identified which need to be addressed; the large number of entries which can be generated without evaluation, identification and resolution of key information to reduce unnecessary effort in concurrent design, the linking of various functional areas, and the perceived lack of efficiency of computer-based systems.

1. Introduction

Previous efforts on a computer supported group design process were aimed at identifying how engineers design and at identifying issues in developing a computer-based design system. This work addressed the design tasks and the communication/management activities of computer assisted group efforts. The results indicated that most of the decisions were made by the manager without explicit rationale and a structured design approach was not followed. A model was sought which would a) provide some structure to the design process, b) document the decision making process, and c) provide an accessible design history so that the design could be revisited for improvement or redesign.

A model which met these criteria was found in the process-based approach to computer-supported Engineering Design [1]. The basic structure of this model is the design matrix which is a structured method based on issues and activities and a deliberation model for capturing the rationale.

A Product Development System was designed to address the many functional areas encountered in product development. For example, some of the matrices in this effort are a Team Matrix, a Customer Matrix, a Product Matrix, a Manufacturing

Matrix, and various product and manufacturing subassembly matrices. The issues vary with each matrix while the activities are the same for all matrices; Generate, Evaluate, and Select.

The product development system is implemented on a small cluster of microcomputers. In addition to the matrices the system includes communication systems for sending/receiving messages, making comments, and management systems for directing the team effort and handling modifications. Entries into the system include text and documents such as drawings or spreadsheets.

2. Background

Group Design System

A group design system (GDS) was developed using a small cluster of microcomputers (seven) and a multi-user data base [4]. The user selects design transactions and enters information using a keyboard or sketch pad. The system incorporates control of the group effort and communications between team members and the manager. The GDS allows for sequencing and identifying the source of the design transactions, provides a framework for categorizing the design transactions, records and stores the content of these transactions, allows for the parallel processing of ideas, and permits controlled flow of information such as message sending. The results of the design process are presented as a timed sequence of interdependent, categorized, design activities.

Three groups of engineering designers were studied using a computer based group design system (GDS). Observations were made regarding the impact of management style (methodology) on the design process, the impact of communications on the overall group efficiency, the sequence of transactions leading to the conceptual design decision, and the involvement of group members. The use of a structured procedure appears to ensure more equal participation, a decision less influenced by individual choice, and reduced authority hierarchy. Two groups did not follow a prescriptive design methodology and the rationale behind the decision making was not clear.

Process-Based Model

The proposed process-based approach to the engineering design of products is based on a design matrix which models the design process and a deliberation model which captures the design rationale [1].

Design Matrix Model

A Design Matrix is used as the basis for the process-based approach to the design of products. The Design Matrix combines the features of design methodology and a design process model. The Design Matrix is a way of organizing the design issues and activities. A schematic for a simplified Design Matrix is shown in Figure1.

In product development the design matrices are more complex (including marketing, manufacturing, maintenance, etc.) and additional design matrices would be included for the subassemblies and components. The format remains the same, only the issues vary. The rows represent the issues to be addressed. In this example the issues are the Problem, Requirements, Functions, Design Concepts, and Design Layout. These issues are addressed by the activities in the columns labeled Generate, Evaluate, and Select.

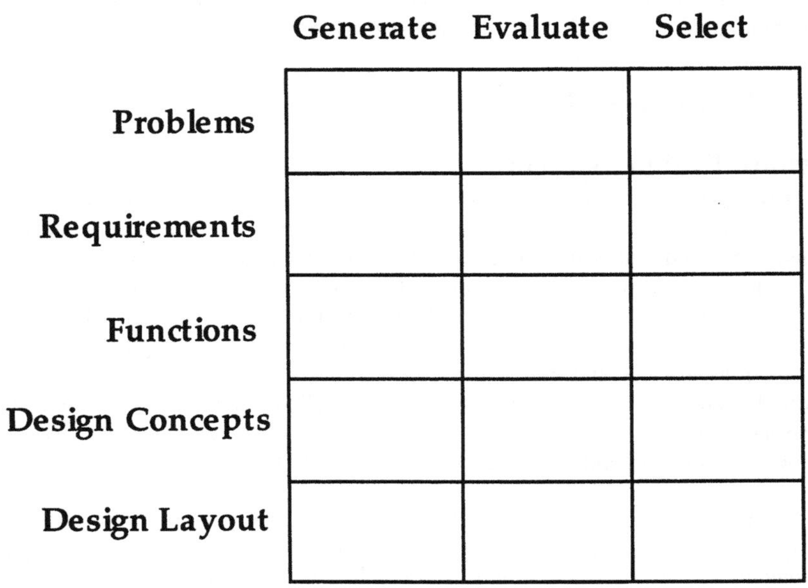

Figure 1: Simplified representation of Design Matrix

Generate - The Generate Activity is used to create proposals for the issue. Modified proposals, while based on previous proposals, are also considered as new proposals.

Evaluate - The evaluation of a proposal is of an absolute comparison of the proposal with the requirements and constraints. An evaluation may result in a proposal being rejected if it violates a requirement or continuing to the selection process if it meets the requirement.

Select - The selection activity is based on a relative comparison of the proposals for each issue to select the best proposal.

Deliberation Model

The evaluate and select activities in the design matrix provide a framework for the deliberation model based on IBIS [8]. The IBIS model is based on the principle that the design process is fundamentally a conversation among the designers, customers, builders, etc. in which they bring their viewpoints and expertise to resolve design issues. The model focuses on the key *Issues* of the design problem.

Each Issue can have many *Positions*. (A Position is a statement which resolves the Issue). Each Position on an Issue may have one or more *Argument* which supports or objects to the position. The system is shown schematically below.

A Position responds to an Issue. Arguments either support or object to a position. Issues may generalize or specialize other issues and may also question or be suggested by other positions and arguments. A discussion may begin with a design concept of how to solve a problem. A person may suggest a position on

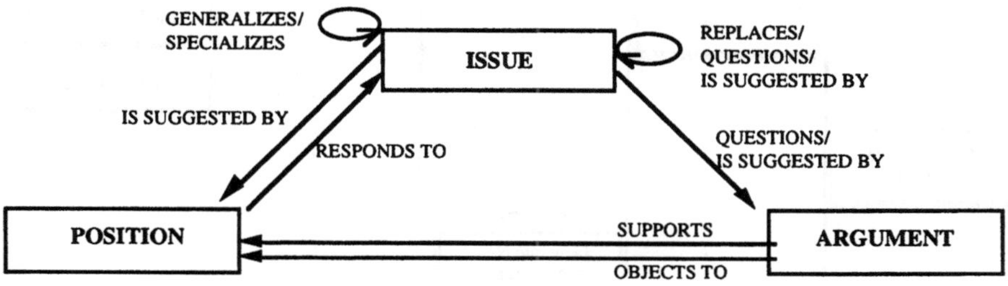

Figure 2: Schematic diagram of IBIS

the concept with supporting arguments. Another person may make arguments supporting or objecting to the position and propose a second concept.

3. Product Development System

The structure of the Product Development System (PDS) is based on the Design Matrix outlined previously. The activities are Generate, Evaluate, and Select. The issues are based on a Matrix Type. For example, a marketing matrix would have issues such as customers, target market, competitors, and market size. There may be more than one matrix of a particular matrix type e.g., there would be many different component matrices with each being a component matrix type.

The PDS system is designed around the Generate Activity. It is this activity which permits the entry of new data into the PDS and it is this entry which is the object of the Evaluate Activity and Select Activity. The Generate Activity may be modified. The Generate Activity is also the subject of questions and comments.

Figure 3 shows the structure of the Product Development System. The user starts with the main control panel where a matrix and issue is selected. Then the user either enters new information (Generate Activity) or reviews existing entries or conducts additional operations on previously Generate Activities. In the section below the details of entering and reviewing the information in the PDS will be outlined and the management and communications associated with the design activities will be discussed.

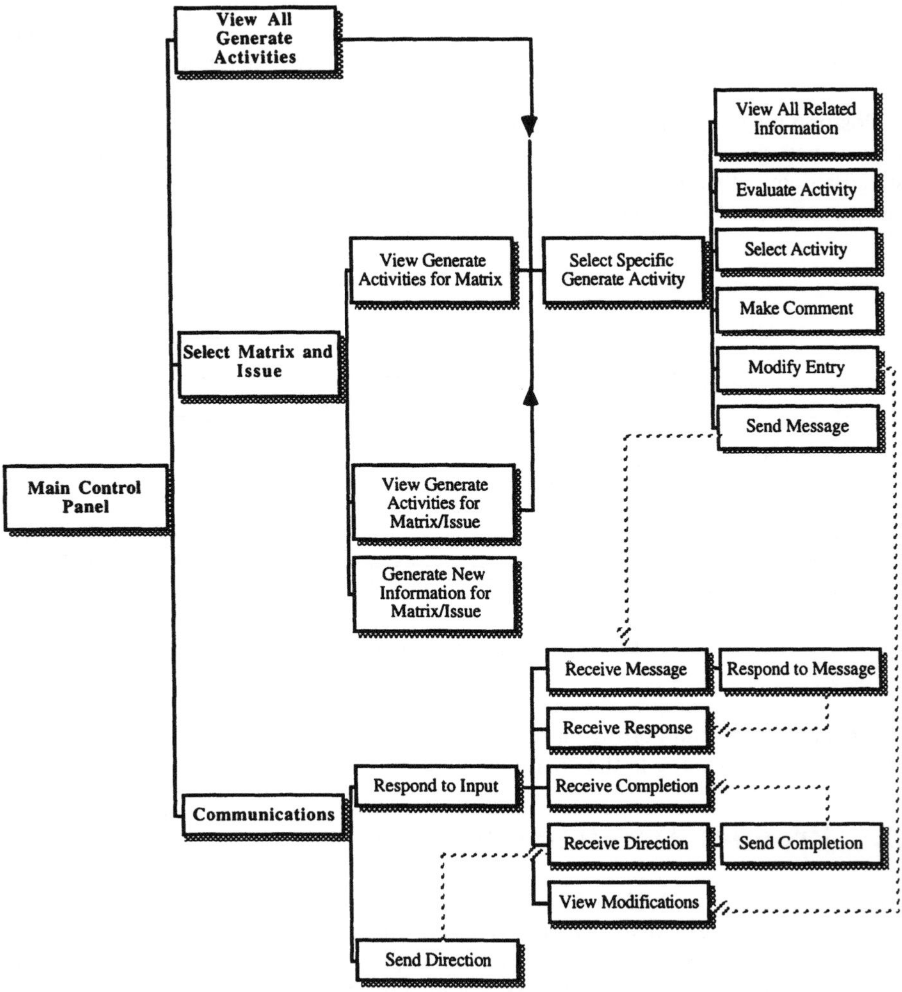

Figure 3: Navigating through the Product Development System

Design Transactions

The transactions can be divided into two types; entries of new information into the system and review and evaluation of information in the system. The basis for all transactions is the Main Control Panel shown in Figure 4.

New Information

To enter new information the designer selects the Matrix Type, the Matrix, and the Issue from the drop down lists and then selects the Generate button. The designer can enter text information describing the new information and/or an application document, e.g. a CAD drawing, which is attached to this activity. When finished, the designer is returned to the control panel.

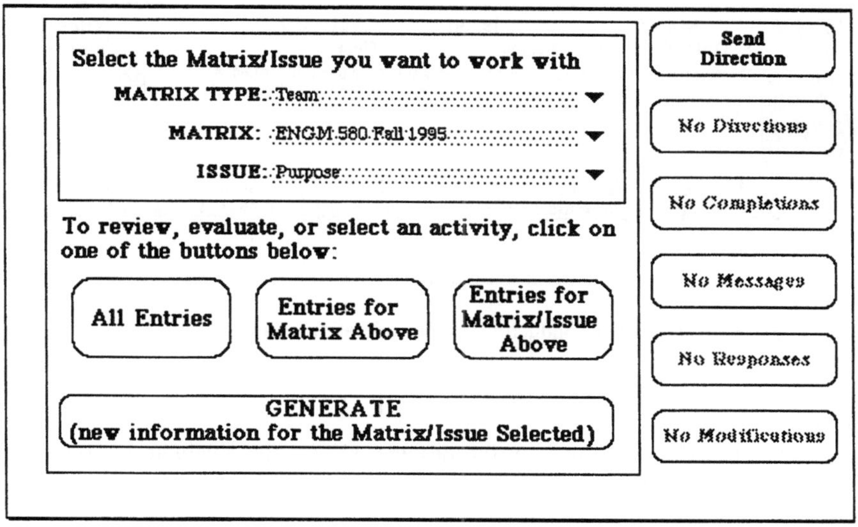

Figure 4: Main Control Panel for Product Development System

Existing Information

To review and analyze existing information the designer can either view all Generate Activities, or select a Matrix Type, Matrix, and Issue and view Generate Activities in a Matrix or for a specific Matrix/Issue. When the designer selects one of the buttons in the Control Panel he/she is presented with the corresponding list of Generate Activities. A sample from the list is shown in Figure 5.

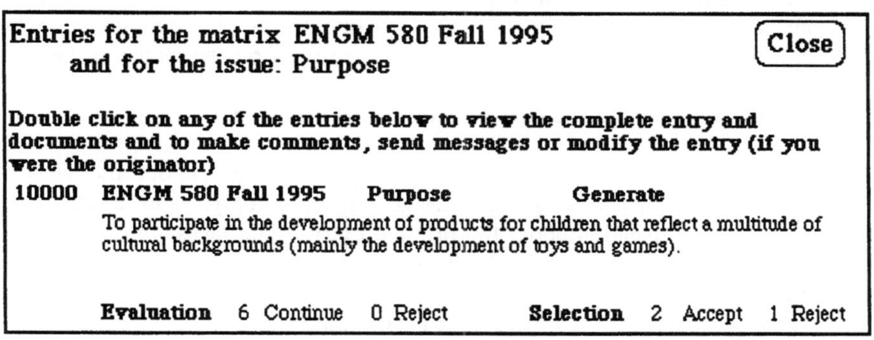

Figure 5: Sample from list of Generate Activities in Product Development System

The information includes the Generate Activity and a summary of the number of evaluations and selections. The designer chooses an activity from the list and is presented with a form to conduct further operations on the Generate Activity. This form is shown in Figure 6.

From this form the designer can evaluate, select, send messages, make comments, modify the entry or view all of the existing information related to the generate entry.

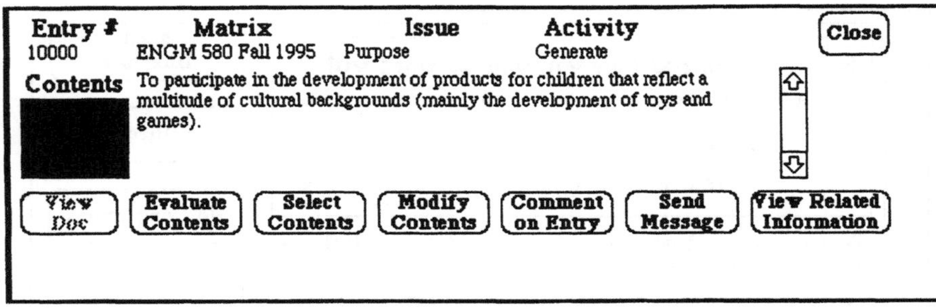

Figure 6: Form for conducting further operations on Generate Activity

Evaluate
To evaluate the Generate Activity the designer selects the Evaluate button and is presented with the Evaluate Activity form. The designer enters the basis of the evaluation, the position, the argument to support the position, and a decision to reject or continue with the entry.

Select
The designer makes a selection by choosing the Select button. The design is presented with the Select Activity form. The Generate Activity is automatically entered and the designer enters the decision being made (accept of reject) and the argument for the decision.

Modify
The team members can also modify a Generate Activity. The team member is presented with a screen which contains the Generate Activity to be modified. When the designer completes the modification the new entry is treated as a new Generate Activity except it is tagged indicating that it is a modification of the previous activity and the team is notified that a modification was made.

The designer can also send messages or make comments about the Generate Activity. The designer sending the message expects a reply. For example, the message may be a question about the original entry. The message is sent to the originator of the generate entry and the originator replies to the message sender. The designer may also send a comment about the Generate Activity. In this case the designer does not expect a reply from the originator.

The Evaluate Activity, the Select Activity, the Modify Activity, the Message and the Comment are all performed on a Generate Activity. By selecting the View Related Information button the designer can view all of these activities in response to a particular Generate Activity.

Management and Communications

The management and communications functions are on the right hand side of the Main Control Panel. Any member of the project team can send a direction to any other member by selecting the Send Direction button. The designer should indicate a date, an activity, and a destination. When a direction is sent to a designer, the control panel for that designer indicates that a direction is waiting. The designer reviews the direction, completes the activity, and indicates that the work is complete. The control panel of the designer who sent the direction indicates that the direction is complete and the designer can view the results.

The communications of sending messages and making comments on specific generate entries are discussed above. The other communication activities are responses to the Send Message, and to the Modify Activity.

When a message is sent by a designer the originator of the Generate Activity will be alerted by the communications part of the main control panel indicating that a message is waiting. The originator of the Generate Activity can then respond to the message. When the originator of the Generate Entry responds to the message, the main control panel of the message sender will indicate that a response is waiting and the designer can view the response to the message. Whenever a Generate Activity is modified all of the designers are notified that a modification has been made.

Other management features are part of the system structure. These management features include: the entries are anonymous, only the project manager can make "permanent" decisions, messages are automatically sent to the originator of the Generate Activity. These decisions are made in the design of the system development and may be modified.

The matrix type, specific matrix and the issues associated with a matrix type are entered by the facilitator or project manager. They can be entered early in the project or as the project progresses. For example, the Marketing Matrix may be entered before the project begins while a specific subassembly matrix may not be determined until the project has developed some concepts.

In addition to the main control panel some additional features are included in the custom menus to aid the designer. The menu includes a list of all of the decisions (permanent selections made by the project manager), a list of the most recent entries (in this case the last week), a list of the entries for the designer using the computer, a list of directions for the designer, a list of directions completed, a list of all the matrices and issues, and a list of the team members and contact information.

4. Experimental

The PDS was implemented in a graduate Product Development course. Eight students were involved in the project for one semester. The class met one night per week. The primary objective of the class was to teach the product development process, introduce the students to the Product Development System, and develop a new product using the system.

Environment

The class was conducted in a computer equipped classroom. The participants sat in two rows of four, facing each other. At one end of the row the server was located as well as a large screen monitor. The monitor could be used for instruction or displaying data for the whole group. A VCR and color scanner which could be used to scan images into the computer were also available.

Project

The Women's Entrepreneurial Center was contacted for possible projects. A project was selected for which a considerable amount of marketing data had already been generated. The problem was aimed at developing multi-cultural toys and/or games. The project was done in cooperation with Cultural Babies, a partnership interested in developing a range of multi-cultural baby products.

Methodology

After an initial lecture on the product development process following Ulrich and Eppinger [7] students were introduced to the design matrix and the Issue Based Information System (IBIS) discussed previously.

The students were instructed in how to browse through the Product Development System by reviewing the Corporate Matrix. Information on the partnership such as the corporate history, corporate objectives, organization, information on the toy industry, corporate resources and corporate strategy had been entered into the system based on information supplied by one of the partners.

Following this lecture/demonstration a lecture was presented on teamwork as outlined in The Wisdom of Teams [3]. A "Team Matrix" was created and used for training the product development team on the use of the PDS. The issues addressed included Purpose, Performance Goals, Accountability, and Approach. The objective here was to develop some basic team consensus on these issues as well as provide training on the system. One goal was to reach consensus on a common purpose which all could agree upon. Another goal was to identify the performance goals which would have to be met in order to achieve the purpose. A third goal was to develop an approach which would apply to each team member and the final goal was to devise a system of accountability to evaluate each team member's performance.

The team members were assigned to specific roles on the team; one project manager, one marketing person who would conduct independent market research and be the team liaison with the partner representing Cultural Babies, three members were assigned to product development, and two members were assigned to process development. The final person on the team was to review the strategy, business plans, and develop the financial justification. The author served as the facilitator and managed the PDS system as needed.

The team members were involved in the product development process from the point of view of the role assigned. The project proceeded with a customer matrix, marketing matrix, product matrix, product subassemblies matrices, manufacturing matrix, and manufacturing subassemblies in addition to the team and corporate matrices described above. The matrices and issues were created by the author as needed.

A weekly questionnaire was developed to monitor the progress of the project and to obtain feedback on the computer-based product development system. A 1-10 scale with the end points represented by the descriptions in parenthesis was used and comments were encouraged. The four questions were: How much progress has the team made towards their objectives? (Little -- A lot) How confident are you that the team will meet its objective? (Not confident -- Very confident) How would you characterize the communication between team members? (Little communication -- A lot of communication) How useful is the computer system in capturing the design data? (Little Use - - Very Useful)

Except for the management and communication tools already discussed, no additional tools were part of the PDS. It was anticipated that additional tools will be added to the system to assist the designers. One of these anticipated tools was a system to aid in the decision making. While the system had a deliberation model built in, the need for a decision system to screen the information generated and a system to make the best choice of several candidates was anticipated. Toward this end additional studies were made on several decision tools. Some of these studies were done using the system to "vote" and some decisions tools, e.g. a matrix evaluation, were performed off line.

5. Results

The results are presented based on the previously outlined experimental methodology. The emphasis is on the process rather than the product. The process will be presented in a sequential fashion; starting with the guided tutorial using the Team Matrix, followed by the effort to identify the customer, the teams independent effort on developing the product, a guided effort to select the requirements and narrow choice of concepts, an exercise to select a concept, and concluding with a review of the rest of the effort. The final phase of the results will include some observations on the designers participation and the results of the weekly questionnaire.

Tutorial - Team Matrix

The objective tutorial was to train the team in the use of the PDS and at the same time generate some of the input which is necessary to form a high functioning team. The effort was guided and one issue was addressed at a time.

Purpose

The first issue to be addressed was the purpose. The objective here is to reach consensus on a single statement of purpose for the team. Fourteen entries and two modifications of entries were developed in the first week. In the second week the emphasis was on evaluation of the expressions of purpose generated in the first week. Four modifications were made in the second week. The statements of purpose received negative evaluations if they were too broad or too narrow, while they received positive evaluations if they agreed with the course description and company objective. Three statements of purpose received favorable consideration during the evaluation activity. After a brief face-to-face team discussion of the three, the team agreed on one statement of purpose.

> "To learn the basics of the product development process by participating in a team effort of developing a commercially feasible product (specifically, multi-cultural toys and games). "

Overall the team produced 14 Generate Activities, 6 Modifications, 72 Evaluate Activities, and 16 Select Activities. The process moved through the Generate, Evaluate, and Select Activities for a single issue (across a row in the matrix) to select one purpose from many for the group. These results are shown graphically in Figure 7. The number and type of the activities are plotted for each time period. Most of the generate activities are on the left in the early time periods and the number of evaluate and select activities increase in the later time periods.

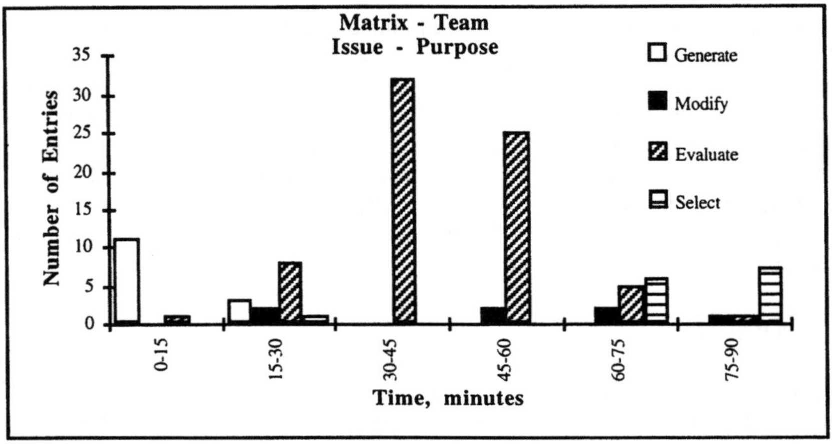

Figure 7: Results of tutorial on Team Matrix/Issue Purpose

Performance Goals

The objective in selecting performance goals was to select a few from many where the objective in selecting the purpose was to select a single purpose from many. The effort on performance goals was completed in the third week. Thirty-three goals were generated and there were 41 evaluations of these goals. In order to reduce the number of goals and select the most important ones the team members were asked to pick their top five goals and assign them a number of 1-5. Based on this "voting" three goals were selected for the team.

> "Identification of product opportunity, the definition of the market segments, and identification of the customers' need"
>
> "Complete a package containing all facets of the project (market, financial factors, design, etc.)."
>
> "Produce a computer model of the selected design and a prototype if appropriate"

The process followed was to Generate, Evaluate, and Select. There were 33 Generate entries, 41 Evaluate, and 40 Select entries. Because few out of many were to be selected a "voting" technique was used to narrow down the performance goals.

Accountability

Having spent most of the time available to work on the PDS for the first three weeks, there was a feeling that the team wanted to "get on with the project" and the issues of accountability and approach were primarily the domain of the instructor. A compromise was for the group to generate some input on accountability and leave it to the instructor for selection and development of the grading scale based on the team input. While there were 23 Generate entries, and one Modification, there was a common thread within the entries and it was easy to select five areas for accountability and grading:

Individual report and team report
Individual progress toward goals
Participation and exchange of information - entries in PDS
Attendance at team meetings (classes)
Peer evaluation - to be developed by instructor.

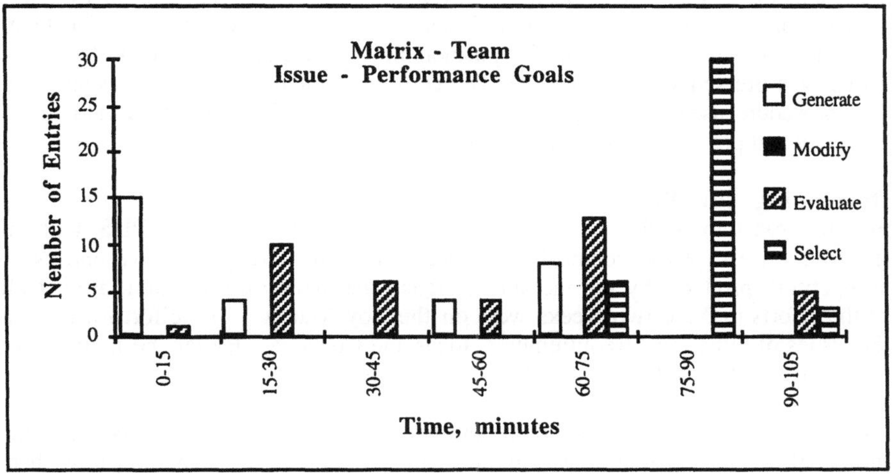

Figure 8: Results of Tutorial for Team Matrix/Performance Goals Issue

Approach

The effort to develop a team approach was to provide some guidelines on how the team should operate to accomplish the goals. These were supplied by the instructor for the reasons mentioned above.

Each individual or sub-group is responsible for completing the assigned goals.
Individuals can and are encouraged to Generate, Evaluate and Select issues in areas other than their own.
When an individual needs information from another area the information should be requested using the Send Direction command.

This completed the effort on developing a team consensus, provided a tutorial for the team, and demonstrated some potential methods for reaching consensus or reducing the number of options. In total there were 81 Generate Activities, 114 Evaluate Activities and 62 Select Activities for the Team Matrix. The efforts now turned to the development of the product.

Customer

It became apparent as the team began to develop the toy or gamè that a key issue was the age of the child. Preliminary research had demonstrated that toys and games are classified by the age of the child. The effort at the end of the third week was aimed at identifying the customer. In the fourth week the emphasis was divided between the Customer Matrix and Toy Matrix.

The three team members from product development, the member responsible for marketing, and the team leader worked on identifying the age of the user. Two positions developed; one group focused on older children over 5 (receive information better, understand multiculturalism, easier to design for), and children less than 5 (desired customers are infants). The position developed in favor to the 5 and under age and two new positions developed based on the research on toys and games which divided this group into 2 and under and 3-5 years old. The team consulted the representative of the partnership. She emphasized that the target group was pre-school. The class reached consensus on infants up to 2 years old. The procedure used by the group was to move from Generate, to Evaluate, to Select as had previously been done in the tutorial on the Team Matrix.

Developing the Product

For the next two weeks the team was allowed to work with the PDS without intervention except for technical assistance as needed. Except for five entries of competitor's products by the member assigned the marketing function, almost all of the efforts in these two weeks were on the Toy Matrix. The effort focused on the issues of Functions, Requirements, Ideas and Concepts. By far the majority of the effort was focused on Ideas (86 out of 118 entries).

Another interesting trend was the predominance of Generate Activity. While in the guided efforts more attention was given to the evaluation and selection, now there were few evaluations and no selections. Of the 118 entries for the Toy Matrix, 99 were Generate Activity, 3 Modify Activity, and only 15 Evaluation Activity.

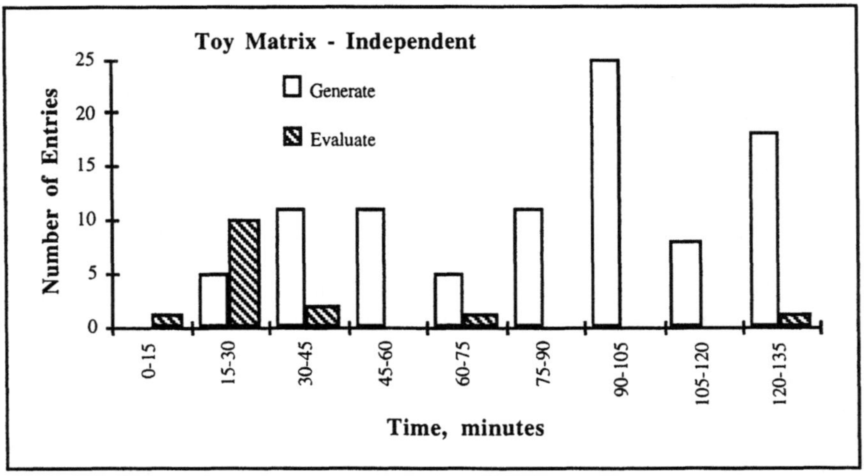

Figure 9: Results of independent effort by design team

Requirements and Concepts
Returning to a guided effort, the team was instructed to look at the ideas and existing concepts and to enter concepts which would best solve the problem. Fifteen new concepts were generated and six evaluations. The team was then instructed to select the first, second and third choice. Based on these selections three different concepts were selected for further evaluation. The concepts were identified as:

Activity Blanket and mobile
Doll with many faces
Shape Puzzle from different cultures

The same procedure was followed to identify six requirements to be used as a basis for selection of the final concept. Requirements were generated (17 and 2 modifications) and evaluated (9). The team was then asked to select five requirements which they felt should be used in the final evaluation. From this selection process the following requirements were selected.

Safety - the toy must comply with the safety and sanitation standards for the target age group, including no small parts or sharp edges and non-toxic paint.
Appeals to senses - needs to appeal to the child's senses, entertain child, be bright and colorful, and utilize muscle activity.
Multicultural - the toy needs to reflect multiple cultures.
Challenging - the toy should challenge the child's growing powers: use developing muscles, agility, knowledge and allow for graduated use for growing minds and bodies.
Cost - relatively inexpensive; the cost of the product needs to be in line with other toys of its type.
Durable - unbreakable, sturdy appearance, resist damage, and washable.

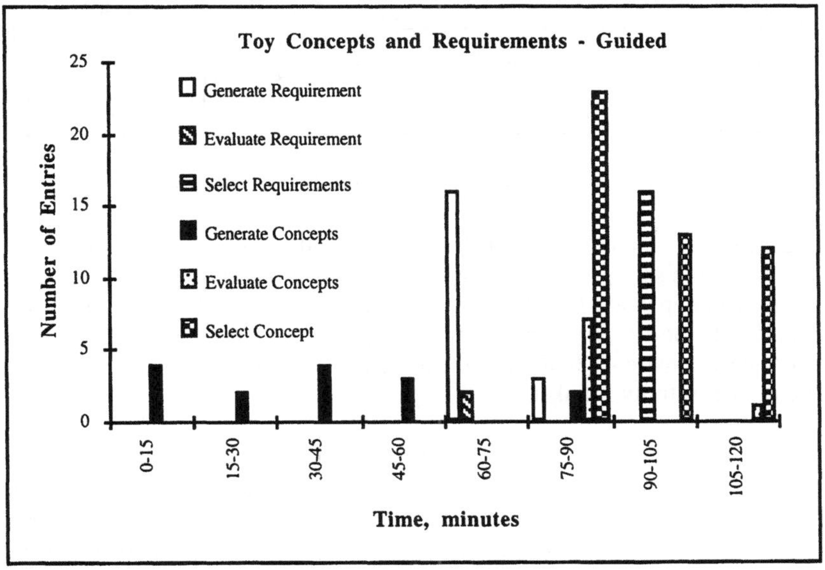

Figure 10: Results of guided effort to select requirements and narrow concepts

Decision Making

A trial was conducted using various decision making methods. These included; selection using the PDS, a matrix evaluation, a questionnaire to be completed by an external group, and a face-to-face discussion.

PDS - The team was asked to evaluate the three concepts based on the six requirements. After completing the evaluation and reviewing all of the evaluations, the team members were asked to select the concept which they felt best fit the criteria. Three team members selected the Doll, two students the Activity Blanket, and two students the Shape Puzzle (one student was absent).

Matrix Evaluation - The team was first asked to rate the relative importance of the requirements. They were then asked to rate the performance of the three concepts. One member selected the Doll, two students the Activity Blanket, and five the Shape Puzzle. When the ratings were averaged for the whole team the concept with the highest rating was the Activity Blanket with the Doll and Shape Puzzle rated about equal. The results are shown in Table 1.

		Concept	
Evaluation	**Doll**	**Activity Blanket**	**Shape Puzzle**
Designer #1	3.93	4.07	**4.60**
Designer #2	3.47	**5.00**	2.47
Designer #3	3.87	4.73	**5.00**
Designer #4	3.93	**4.47**	3.47
Designer #5	**4.53**	4.47	3.13
Designer #6	3.13	3.53	**3.60**
Designer #7	3.27	2.93	**3.93**
Designer #8	4.13	3.80	**4.33**
Class Average	3.73	**4.06**	3.74
Ind. Selection	1	2	**5**

Table 1: Results of Matrix Evaluation by team members

Questionnaire - The third method involved evaluation by "customers" from outside the class. For this purpose a class on entrepreneurship was used (our representative to the partnership was taking this class). The nineteen members of this class were provided with the six requirements and a description of the three concepts. Based on this questionnaire 42% of the students selected the Shape Puzzle, 32% the Doll, and 26% the Activity Blanket.

Face-to-Face - Following these selections the team was asked to discuss the results and make a selection in a face-to-face meeting. The team quickly reached a consensus. The team selected the Activity Blanket. One of the main factors from the video tape seemed to be that the Activity Blanket was the selection of our representative to the partnership (considered by most of the team to be the team's customer).

Following the four decision making techniques a questionnaire was presented to the team. The questions and responses follow:

Which method had the most impact on your decision?

Method	Score	Percentage
Decision Matrix	30	28.6%
Face to Face Discussion	32	21.9%
Computer Ealuation/Selection	21	20.0%
"Customer" Questionnaire	16	15.2%
Company Representaive's Preference	15	14.3%

Table 2: Results of questionnaire on decision techniques

What did they like and dislike about the decision making process?

<u>Likes</u>

Method of using computer in decision making
Learn different method yield different results
Gives opportunity to evaluate different options

<u>Dislikes</u>

Computerized
Letting customers evaluate and make decisions when they do not clearly understand the concepts
Not enough time
Not enough participants
Some methods lack customer input

Why did the team reach consensus quickly in the meeting?

Very good technical support (computer)
All choices were good
Originally, participants liked blanket idea
Company representative's preference led to consensus
Average class choice from Matrix Evaluation calculation
Ideas were similar
Time is quickly running out

Balance of Project

For the balance of the project the majority of the effort was spent on the Generate Activity for the Blanket, Figures, and Mobile Subassembly matrices. Effort was also spent on the Blanket and Toy Manufacturing Subassembly matrices. The final description of the toy including sketches was completed. The last week was spent primarily off the computer planning the final report and presentation. Some entries were made to summarize the decisions. These efforts were similar to the previously described independent effort on the Toy Matrix. A total of 122 entries were made. Of these entries 103 were Generate Activity. There was little attempt to evaluate or select the entries.

Weekly Questionnaire

The overall results of the weekly questionnaire are shown in Figure 11. The rating on the computer system and confidence in reaching the project goals remained fairly constant during the duration of the project. The rating on the computer system started at a 7 and had a range from 6.4 to 8.3. The average was 7.4 and the final rating was 7.9. The confidence rating started at 6.6 and had a range from 6.6 to 8.4 and the final rating was 8.4.

The ratings on the progress and communication started lower and increased in rating as the project progressed. The rating on communication started at 3.1 and had a range of 3.1 to 8.1. The average was 6.4 and the final rating was 8.4. The ratings on progress started at 1.9 and had a range from 1.9 to 8.4. The average was 6.0 and the final rating was 8.4.

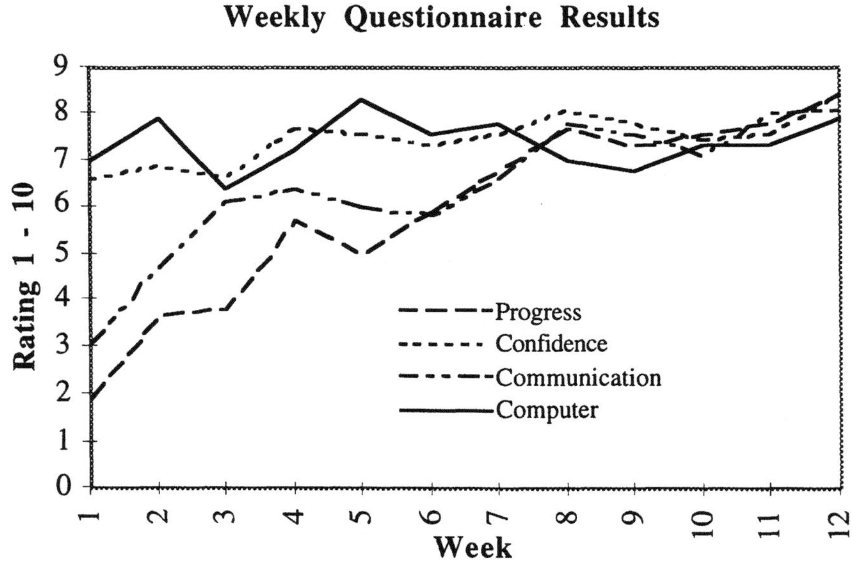

Figure 11: Results of weekly questionnaire on team effort and computer system

Representative comments on the computer system are shown in Table 3 . The table is divided into comments which seem to support the system and comments which indicated deficiencies.

Participation

The total number of entries for the individual designers was tabulated. The number ranged from 51 to 119. The team member with the 51 entries missed four of the classes. The range for the rest of the participants was 74 - 119. The average for these seven participants is 95 entries and the standard deviation is 16. Compared to this relatively uniform distribution on the computer system, the face-to-face session to select the concept was dominated by three or four team members.

Supportive Statements	Deficiencies
Supports all eight terminals in real time	Operational problems
Good job of recording data	Too many entries to evaluate
Everyone has equal amount of input	Questions the quality of decision
Decision without emotion	Time lag in communication
Updates information in real time	Difficult to reverse entry
Captures ideas uninterrupted	Not sure where to enter information
Members work at same time	Hard to see total picture
Direction is clear	Good ideas may be terminated
Keeps track of all ideas	Sometimes discussion helpful
Look at previous work	Loss of efficiency
Records decisions/selection	Overload of information
Capture and organize ideas	Need access from home
Opportunity to provide input in other areas	Difficult to see overall direction of project
Provides information between groups	Captures what is written - may be difficult to express ideas in writing
Good as final input to discussion	
Standard design helps exchange of data	
Great flexibility	
Effective in capturing information	

Table 3: Comments on computer system from weekly questionnaire

Accomplishment vs. Goals

The team had developed three performance goals. The team relied mostly on the information supplied at the beginning of the project for market size. They did identify a specific customer for the toy and thereby some of the parameters for the toy. Competitor's products were identified. The team identified a conceptual design for the product and manufacturing methods for the subassemblies. The main part of the design package which was lacking was the financial analysis. A three dimensional computer model of the design was completed which could be viewed from different angles. Overall, the team accomplished the objective and performance goals they had developed.

6. Discussion of Results

The discussion below is aimed at examining the lessons learned from the semester long use of the PDS in a graduate product development class. Some of the problems and benefits identified may be particular to the system as described previously and some may relate to computer-based efforts in general.

Tutorial - Team Matrix

The system design, shown in Figure 3, where the designer starts by selecting a matrix and issue and then either selects a new Generate Activity or reviews existing information seems to work well. The team did not have a problem learning to navigate the system. The guided effort on the Team Matrix to teach the system was successful both as a tutorial and as a means of establishing consensus on the team purpose and performance goals. During this training, one issue at a time was addressed by all team members. When working on the Team Purpose the effort was aimed at reaching consensus on one purpose out of many. This was accomplished by narrowing down the choices using a review of the evaluations, suggesting possible statements which seemed to capture the sense of the team, and reaching consensus by discussion.

The objective for the Team Performance Goals was to select several out of many and this was accomplished using a "voting" system where the team members were asked to select their top 5 choices, the results were tabulated and the three with the highest number of points were selected. This effort demonstrated the need to reduce the number of candidates for selection.

The use of a team increases the number of alternatives and the use of computer supported teams increases the number of alternatives over face-to-face meetings. On the computer-based system everyone can "talk" at the same time. In contrast to the face-to face meeting where alternatives are evaluated as presented, many alternatives are generated before the evaluation process begins using the computer-based system. Over 20 Generate/Modify entries had to be reduced to a single purpose for the group.

Customer

After the tutorial on the Team Matrix the team turned their efforts to defining the customers. While five of the team members focused on the age of the child, three members began to work on the product matrix without knowing the target age. Each seemed to work with an assumed age. Evaluations were made of ideas based on age appropriateness. In some cases the assumed age was specified; "..because the toy is for a baby" or ".."assuming the child will be old enough to use a computer".

While the system allows for concurrent, individual effort it also allows individuals to go in different directions and some of this effort may be not be effective.

Developing the Product

Eighty-six ideas were generated with little evaluation. The comments from the team on the questionnaire suggested some frustration with the amount of information being generated. The evaluation of the system began to produce comments like "cumbersome", "hard to enter every thought that you have", and "overload of information". This large number of unevaluated entries makes the reduction process difficult and has a significant impact on the solution space. Ehrenspiel and Dylla [2] studied individual designers. They suggested that many use a corrective solution generation where an initial solution is generated and followed until a problem is found when it is corrected or replaced. They recommended the production of several solutions and while this may lead to new interesting solutions it increases the difficulty and cost of the evaluation process. They suggested that it was bad methodology to generate a "senseless, non-reducible flood of variants". The desired approach, also outlined in Pugh [6] is controlled convergence. When the potential for a large numbers of generated alternatives is magnified in a computer-based team development system, the concept of controlled convergence becomes a necessity.

Requirements and Concepts

A guided effort was made to aid in the convergence of the requirements and concepts in the next phase. The team generated 15 concepts which would provide the desired toy for infants to two year olds. Using the voting technique, the team reduced the potential concepts to three; a reasonable number to deal with in a selection process. Similarly, the requirements were narrowed to six for use in the selection process. This led to the decision trials aimed at using and evaluating several decision making procedures to provide guidance for future system development.

Decision Making Procedures

As outlined previously four procedures were used: computer-based selection, a decision matrix, a "customer" questionnaire, and a face-to-face discussion.

Using the computer-based PDS, where the students were asked to evaluate the three alternatives based on the six requirements and select the alternatives each preferred, three selected the Doll, two the Activity Blanket, and two the Shape Puzzle. For the matrix evaluation each was asked to weight the requirements and rate the alternatives. If the individual efforts are considered five selected the Shape Puzzle, two the Activity Blanket, and one the Doll. If the average weighting and ratings are used the Activity Blanket is the choice of the team. In the questionnaire done outside of the class eight "customers" selected the Shape Puzzle, six selected the Doll and five the Activity Blanket. Up to this point the selection seemed to focus on the Shape Puzzle. In the face-to-face discussion the Activity Blanket received the most attention. The team was made aware that it was the choice of the partnership representative who many viewed as the real customer of the design. This seemed to have some impact on the decision. The group reached consensus relatively easily and in a short time. The conversation was dominated by about half of the team.

The team was given a questionnaire to gain more understanding of the decision making process. The team preferred the matrix evaluation. While they used the results of the "customer" questionnaire as input they were not sure the "customers" were familiar enough with the development and relied heavily upon the response of the partnership representative, which favored the activity blanket. Based on the response to the questionnaire the team seemed to think that all of the concepts were good, the concepts were similar and time was running out.

Weekly Questionnaire

The weekly questionnaire was helpful in evaluating the system and suggesting intervention by the facilitator. In general the comments suggest that the system possessed the benefits normally attributed to computer-based support systems: simultaneous processing of information, equality of input, documentation of all information, and information available for review. In addition the matrix design provided a common structure for many functional activities, permitted team members to review and input into other functional areas. Several of the problems with computer-based support systems were also identified including the difficulties in typing in all information, the time lag in communications, the loss of efficiency and the overload of information. Some of the problems may be particular to the system or represent the limited use: difficulty in reversing entries, difficulty seeing the "total picture", and the lack of access outside the classroom.

During the process some modifications were made to ease these problems. A pull down menu was added where the team members could review the decisions made by the team. The menu also included recent entries (the last week) and entries made just by the team member. The listing of the final decisions was especially helpful to the team members.

7. Conclusions and Recommendations

The PDS system, based on the Design Matrix was a successful system for organizing product development information from a variety of functional areas and at different levels of detail. The deliberation model, used in the evaluate and select activities, captured the rationale used in the decision making.

In some cases an unwieldy number of inputs were generated without evaluation or selection. It is necessary to develop the procedures and means to both reduce the number of alternatives to a manageable number and the means to move towards consensus. Tools to ensure controlled convergence are necessary.

It is necessary to identify key information needs and plan for the timely decision making on this information. For example, the age of the child was a key piece of information. Some designers developed ideas and concepts without this information and some of this effort was wasted once the age was selected. The planning is especially necessary for concurrent design efforts where the overlapping of activities by functional areas requires that tentative information be utilized while the risks of unnecessary work is minimized.

Another issue is linking the efforts from the various functional areas. The results of the issues, especially requirements, selected by one functional area will have an impact on other areas.

The perceived lack of efficiency of the computer-based system is an important issue. While technology such as voice entry of data may address the typing deficiencies, the positive benefits of face-to-face discussion should also be recognized. There is some indication in the study that of electronic data collection/evaluation in combination with targeted face-to-face meetings may be effective.

A management approach based on managing the process instead of managing the budgets and schedules needs to be developed. In the current scheme, it is the responsibility of the project manager to make the final (permanent) evaluations and selections to reject or accept a proposal. A new scheme or a more trained, active manager will be required to enable this system to be successful.

Four roles are suggested for the manager/facilitator:

- Negotiate contracts such as was done in the team matrix.
- Promote rational decision making as was done in the evaluation and selection. A more active role by the facilitator to make permanent acceptance/rejection is necessary.
- Manage the process to maintain a manageable number of concepts through controlled convergence.
- Manage information to minimize risk and unnecessary work in concurrent design.

This study describes a framework for successful computer supported, co-operative product development using a process based approach and identifies areas which must be addressed for an effective and efficient system.

8. References

1. Blessing L T M 1994 Process-based approach to computer supported engineering design, PhD thesis, University of Twente, the Netherlands

2. Ehrenspiel K, and Dylla N 1993 Experimental investigation of designers' thinking methods and design procedures. *Journal of Engineering Design*, Vol. 4, No. 3

3. Katzenbach, Smith 1993 *The Wisdom of Teams*. Harvard University Press

4. McMahon E H 1991 Group design system. In: *Proceedings of ICED 91*, Edited by V. Hubka Zurich, Switzerland August 27-291991 WDK 20 Heurista Zurich, Switzerland

5. McMahon, E.H., 1995, A process-based approach to product development. In: *Proceedings of ICED 95*. Edited by V. Hubka Praha, The Czech Republic August 22-24,1995 WDK 23 Heurista Zurich, Switzerland

6. Pugh, Stuart 1990 *Total Design*. Addison-Wesley Publishing Company Wokingham, England

7. Ulrich and Eppinger 1995 *Product Design and Development*. McGraw Hill New York, New York.

8. Yankemovic K.C Burgess and Conklin E. Jeffrey 1990 Report on a development project - use of and issue-based information system. *Proceedings of the Conference on Computer Supported Cooperative Work* October 7-10 Los Angeles, CA Assoc. for Computing Machinery, New York

Competitive Industrial Product Development Needs Multi-Disciplinary Knowledge Acquisition

Margareta Norell
Integrated Product Development, Dept of Machine Design, KTH
S-100 44 Stockholm, Sweden

Abstract

The objective with this paper is to discuss the potential in use of multidisciplinary methods in Engineering Design research. The thoughts are based on experiences from several investigations concerning efficiency in product development in industry performed within an interdisciplinary research programme. The investigations include technical, organisational as well as human aspects. A selection of results are presented from studies concerning impact on concurrency and efficiency from implementation and use of development tools and changed work methods in product development processes. The results show implementation of new methodologies and integrated product development to demonstrate high potential for more competitive development processes with shorter lead time and higher over-all quality. Of major importance for successful implementations are simultaneous focus on process efficiency and working conditions, which require a multidisciplinary approach.

1. Introduction

Changing industrial conditions put focus on the adaptation of the industrial organisation. Increased competition contributes to the demands for higher efficiency and productivity in the over-all process. The product development process has been found to play an important role and has consequently received an increased interest. Key factors in the process are time, quality and cost. There is a great challenge in the search of knowledge concerning long-term product development process efficiency.

Product development in this paper refers to the whole process of product realisation, though focusing on synthesis and analysis of new product concepts. Product development includes complex combinations of for example technical, economical, social and marketing activities. An important model considering the process is the Integrated Product Development (IPU) model, which has its origin from Scandinavian works [1] and [2], and has been further developed and applied [3].

Short lead time is a major competitive factor: parallel activities instead of sequential in the process is one way to reduce the calendar time. Efficient communication and close co-operation between different groups and functions are essential. The necessity to consider tasks early in the process, including human aspects, has been increasingly recognised during the last years. It is of major importance to use and develop the competence of the people acting in the development process, supporting both process efficiency and personal satisfaction [4]. Another factor of great importance is decisions in early phases of the development process, these decisions should include considerations concerning for example market needs, quality, manufacturing, and life cycle costs. Japanese companies have shown to be highly competitive in that respect by "doing right from the start" [5].

The difficulties with early considerations are obvious - several development tools have been introduced and used with the purpose to improve efficiency in different phases of the development process. Some of these tools have considerably affected the development process, by making shorter activity loops possible. However, it is not clarified to what extent these tools have influenced the process efficiency which is not necessarily equal to short activity loops [6]. There is still need for methods to evaluate impact on over-all quality of the process.

The concept of Concurrent Engineering (CE) shows a lot of similarities with the concept of IPU, concerning product development work carried out in parallel processes and with a high degree of co-operation between different domains.

Three ingredients are of major importance when discussing implementation of Concurrent Engineering or Integrated Product Development [7]:

- *Work methods* - Organisation and management supporting integration.
- *Support tools* - Use of efficient tools for support in product development.
- *Information systems* - Use of relevant information systems

The work methods are vital - an integrated working methodology requires an organisation with involved management, motivating co-operation between different domains. The organisation of development work in cross-functional teams is recognised as very fruitful. The project manager has an important role to establish the same objectives for everyone in the project [8].

2. Interdisciplinary Studies

An interdisciplinary research programme in co-operation between researchers from engineering design and work psychology was initiated in 1988. The objective of the programme has been to develop knowledge concerning efficiency in product development processes regarding both technical and organisational questions. Several investigations have been performed within the programme, this paper presents three studies concerning development tools and one concerning feasibility studies in product development work.

The purpose of the studies concerning development tools was to investigate and analyse factors of importance in practical use in industrial product development. Areas of questions were:

- How does the development tool affect project efficiency, product, and co-operation in the process?
- Are there any effects on learning and competence development observed to be dependent on work with development tools?
- What characterises successful implementation and use of the development tool?

The purpose of the investigation concerning feasibility studies was to find key factors for successful early decisions. Question areas were:

- How are the feasibility studies organised and performed? Are there strategic considerations made?
- Which factors are perceived as important for an efficient feasibility study?

2.1 Studied development tools

Development tools are here defined as manual or computer based systematic methods or frameworks which have the potential to increase efficiency in one or several phases of the product development process. Implementation and use of five development tools were studied with the purpose to find factors that affect process efficiency.

The tools studied are chosen with the requirements that they should:

- Address a specific problem/phase in product development.
- Be used by several functions and persons in several industries.
- Have a potential to be a "bridge builder" - a forum for co-operation.

Based on these criteria the following development tools were studied:

DFA - Design for Assembly, according to [9]. DFA is used to point out parts in the design or concept which need further attention for assembly cost reasons.

FMEA - Failure Mode and Effects Analysis is a method which is used to find and judge potential sources of error in products or manufacturing processes, FMEA also includes judgements of consequences [10] .

QFD - Quality Function Deployment is a method to translate the customers needs to technical requirements, and is used to form well rooted requirement specifications [11].

LCA - Life Cycle Assessment is a method for identifying the total environmental impact of the product during the entire life cycle. EPS - Environmental Priority Strategies is an evaluation system in LCA [12].

PDM - Product Data Management systems are computer based structures which support the management of product data, the product development process, the product realisation and the documentation [13].

2.2 Feasibility studies

The feasibility studies in the investigation are defined as the activities between a first, preliminary, market specification and the concluded product specification. The concluded product specification and the detailed plan for the following product development project, form the base for the decision of "go-no go".

The investigation concerning feasibility studies in product development was based on the experience that new strategies for extensive feasibility studies were discussed and implemented in some industrial companies. The study was performed in two industries with expressed intentions of change in feasibility study work [14].

3. Research Method

The research method used in the studies is an adapted version of Grounded Theory [15]. The defined questions were investigated through interviews with people in practical work in product development projects. The interviews were all semi-structural according to pre-developed interview guides, the guides were tested and discussed in the research group before use. The structure of the interview guides opened for the interview persons to influence the discussion and to allocate relatively more time to the most important questions for him/her. Each interview was going on for about one hour, they were all tape recorded and afterwards transcribed and coded into categories by at least two researchers independently. The coded protocols were then condensed and analysed and provided the basis for the reported findings. All results were discussed with the participating companies.

Different companies have participated in the different investigations, totally twelve companies have been involved in the presented studies. Normally the companies for one specific study are chosen according to a well worked through list of criteria, which differs between the studies.

All twelve included companies are developing and manufacturing their own products, they all have a large share of export sales, the products are mechanical and electromechanical. A number of product development projects have been included in the studies and the data have been collected by the researchers mainly by interviews.

The interview persons were operationally involved in product development work, representing several functions. The volume of interviews differs between the studies, for DFA, FMEA and QFD about 30 experienced users from different functions were interviewed. The study of LCA includes 16 interview persons and the PDM study 20 persons. The feasibility study investigation included 27 interviews at different stages in the process. On an average 50 - 55 % of the interviewees were representing design, about 10 % were managers.

4. Results

4.1 Development tools

4.1.1 Effects on project efficiency, process and co-operation

The studied development tools were in general introduced to support a specific technical factor in the development process. Many of the implementations have, however, shown to also affect other factors in the process, affecting the general understanding of the demands. These factors are viewed as important as the more technical effects.

All the studied tools have the potential to support designers and others individually. To support the efficiency of the process, the support tools should not only impact individuals, but also constitute a platform for communication. The studied tools are all perceived as very efficient and relevant in the domain they address. Furthermore the results from the studies show that they all can improve the interaction between people involved in a product development project, and be a relevant base for team-building.

Applications of DFA, FMEA and QFD are in the studied applications in general shortening the total development time [16].

Use of LCA and EPS in product development give the possibility to make environmentally sound choices in early phases, which could decrease the numbers of late changes and therefore affect the lead-time and process convergence. Several interviewees reported that EPS may encourage co-operation particularly with sub-suppliers. [17].

PDM systems are perceived as having the potential to be an important support in the continuously more complex structure of data management. The product development process should be analysed and - if needed - improved before the implementation of a PDM system. PDM systems have the potential to increase the co-operation especially between partners with different geographical locations. [18]

DFA, FMEA, QFD, LCA and EPS are all contributing to the adding of more knowledge to the work in an early stage of the product development process according to the interviewees. The general recommendation is to use the methods as early as possible in the product development process, to decrease the number of errors and changes during the process. PDM is today mainly used in the detailed design, but the intention is to broaden the use also to the earlier phases.

4.1.2 Impact on user - learning and competence development

No design problem solving functions are included in any of the studied tools, they can be characterised as "problem pointing" and leaves the creative work to the user. The risk is therefore minimal that the methods will cause an impoverishing effect on the work in the product development process.

All the interviewed persons in the studies concerning DFA, FMEA, QFD, LCA and EPS have the opinion that the usage increases knowledge and competence. The tools are contributing to the learning in both depth and width in the addressed domains. All the studied development tools, including PDM systems, contribute to the process orientation and a holistic view on the product development process.

4.1.3 Implementation and use

Experiences have shown that a high degree of awareness is required to reach success in making the product development process more efficient. If the process of increasing efficiency is being strengthened with the help of development tools, a conscious strategy is also needed to succeed in the implementation of tools. One important factor is to analyse the "as-is" situation in the product development process before changes in terms of new tools are made.

The studies clearly show that an analysis after the first period of use is of major importance. The first attempt shows rarely long term effects without analysis. The analysis should be followed by a decision concerning continuous use of the tool or not. Resources should be allocated to adaptation, education, support and guidance for successful application to the internal organisation of processes.

4.2 Feasibility studies in product development

4.2.1 Strategies and organisation

The investigated feasibility studies were the first generation of a changed strategy with the purpose of allocating more resources to the earliest phases. The strategy could be formulated: By putting more resources in terms of increased time, travels, market analyses and by dedicating several competencies into the feasibility study the over all product development process will win in efficiency. The specification will be better anchored, and better fulfil market demands, the development project will be more predictable, shorter and more convergent.

The feasibility studies were organised and performed as team projects. The projects required input from many disciplines, in a mixture unique to the considered product. Co-operation was facilitated by using project matrix organisation, where a dedicated project manager co-ordinated and participants in the multidisciplinary team. The goals of the feasibility studies were committed product specification, project specification and a prototype. The committed result supported the establishing of a realistic plan for the following product development project and avoidance of late and expensive changes. The organisation model also made it possible to create off-the-shelf variants of components, which contributed to continuous improvements of the products.

4.2.2 Important factors

Detailed planning of the feasibility study, with strict time schedule, should be avoided to allow a creative climate, but it is still important to formulate a common ultimate goal. The learning participation from manufacturing people is important, these representatives report difficulties with abstraction. The working conditions are considered very important, and should generate creative, open-minded, stimulating and efficient product development work. More resources and higher degree of autonomy in defined expanded feasibility studies increase the process concurrency and predictability in the development projects. [14]

5. Conclusions

Several of the studied tools, and many others, can be a powerful support in striving for integrated product development if they are implemented with regards to a holistic approach on development processes and work. Both implementation of new tools and new strategies for feasibility study work are recommended to start the change process. The change process must then be treated as a process of continuos improvement - with all the ingredients of observation - analysis - reflection - change [19].

A clear finding in the studies is that competitive development work requires motivated persons in the process. Basic psychological needs for human beings are perceived autonomy, possibilities for a comprehensive view of the task and opportunity for development and learning.

Some basic factors concerning working conditions can be formulated for efficient industrial product development processes:

- True multidisciplinary co-operation is of major importance, and can not be obtained without support in the organisation and management.
- Development methods and tools, which are not perceived as efficient by the user, will not be optimally used in long term.

- Clear, commonly formulated objectives are important, and contribute to the reduction of functional and prestige-related barriers.
- People with authority to take responsibility are stimulated, motivated and report better results.

6. Discussion

The investigations summarised in this paper were performed with an interview methodology (described under Research method). The validity, the strengths and the weaknesses with the method can be further discussed.

One major characteristic with the method is the purpose to collect data with as little subjective impact from the researchers as possible. This puts emphasis on the ambition and competence of the researchers to remain objective, which eventually need special education or training. It is strongly recommended to co-operate with human behaviourists concerning these issues.

If the interview guide and the researcher are objective the method has the potential to mirror the "real situation" in a unique way. The author knows no better method for collecting data and mapping industrial experiences, forming the "as-is" situation.

An often reported effect of the interview series is the starting of a reflection process within the companies, where the participants after the interviews have begun to think and discuss different possibilities to develop their own processes. Consequently one spin-off effect could be a learning network, where the change of experiences and thoughts can contribute to increased use of resources. The research could eventually therefore also be seen as action research [20]. The process of developed understanding in the mind of researchers, interview persons, and in industrial organisations is of special interest in a dynamic competition, perhaps the common learning and the "learning to learn" is "the thing" with this kind of research? [21]

Some other characteristics which should be pointed out are that these types of investigations must be performed in research teams - preferably including several research disciplines and personality types (like efficient product development projects!).

There are also several weaknesses with the interview methods, some of them mentioned here: The data collection and the analysis work is not computable in a very efficient way and statistical methods cannot be used. The methods demand a lot of time-consuming work, only relatively small populations can be involved. The results cannot be mathematical measured, which cause problems for some of us engineers.

Competitive product development is a challenging issue, including questions concerning several disciplines. Research and industrial search of knowledge demand contributions from both technical and human behaviour competencies.

There are no turn key-solutions to be found, but there can be found clear indications of fruitful first steps and directions.

The impact of individual and organisational learning should not be underestimated as a factor for increased competitiveness - the process of change of methodologies in product development should be organised to stimulate both individual and organisational learning.

The competitive product development is a multidisciplinary task needing multidisciplinary knowledge acquisition!

7. Acknowledgements

The simultaneous focusing on both technical processes and individual demands need different kinds of knowledge to be mixed in research. This has been made possible by the double funding from The Swedish National Board for Technical Development (Nutek) and The Swedish Council for Work Life Research. The author specially thanks Dr Svante Hovmark for years of fruitful co-operation.

8. References

[1] Olsson F 1976 Systematisk konstruktion (Systematic Design), in Swedish. PhD thesis, Department of Machine Design, Lunds Tekniska Högskola

[2] Andreasen M.M 1983 *Integrated Product Development - a new framework for methodical design.* Proceedings of ICED 83, Heurista, Zürich

[3] Andreasen M.M, Hein L 1987 *Integrated Product Development.* IFS Publications Ltd, UK, Springer Verlag

[4] Senge P 1990 *The Fifth Discipline - the Act and Practice of a Learning Organisation.* Century Publishing Comp

[5] Womack J P,. Jones D T, Roos D 1991 *The machine that changed the world.* New York: Harper Perennial

[6] Larsson G 1996 Han leder Saabs andra revolution (Managing the second revolution of Saab), in Swedish. *Dagens Industri* 5 mars 1996

[7] Norell M 1992 Stödmetoder och samverkan i produktutveckling (Advisory Tools and Co-operation in Product Development), in Swedish. PhD thesis, TRITA-MAE-1992:7, Department of Machine Elements, The Royal Institute of Technology, Stockholm

[8] Hovmark S, Nordqvist S 1993 *Change to more efficient product development: Working methods and collaboration in five product development projects* (in Swedish). Rapporter, nr 72. Department of psychology, Stockholm university

[9] Boothroyd G, Dewhurst P 1989 *Product Design for Assembly.* Boothroyd Dewhurst, Inc,Wakefield, RI.

[10] IEC Standard 1985 *Analysis techniques for system reliability - Procedure for failure mode and effects analysis (FMEA).* Bureau Central de la Commission Electrotechnique Internationale, Genève, Suisse

[11] Sullivan L P 1986 Quality Function Deployment. *Quality Progress,* June 1986.

[12] Ryding S-O, Steen B 1991 *The EPS SYSTEM, a PC-based system for development and application of environmental priority strategies in product design - from cradle to grave.* Swedish Environmental Research Institute, IVL B 1002, Gothenburg.

[13] CIM-data, Inc. 1992 *Product Data Management: A Technology Guide*

[14] Frisk E, Nordqvist S, Norell M 1995 Concurrent Engineering - Strategies for Feasibility Studies in Product Development. *Proceedings of ICED 95,* Prague

[15] Pidgeon NF, Turner BA, Blockley D I 1991 The use of Grounded Theory for conceptual analysis in knowledge elicitation. *International Journal of Man-Machine Studies,* 32, 327-340

[16] Norell M 1993 The use of DFA, FMEA and QFD as tools for Concurrent Engineering in Product Development Processes. *Proceedings of ICED 93,* The Hague

[17] Ritzén S, Hakelius C, Norell M 1996 Life-Cycle Assessment, implementation and use in Swedish industry. *Proceedings NordDesign,* Helsinki

[18] Hakelius C, Sellgren U 1996 *PDM projects in Swedish industry - experiences from six engineering and manufacturing companies.* TRITA - MMK 1996:6, Department of Machine Design, The Royal Institute of Technology, Stockholm

[19] Forslin J, Thulestedt B-M 1993 Lärande organisation. Att utveckla kompetens tillsammans (*Learning organisation. To develop competence in common),* in Swedish. Publica, Stockholm

[20] Schön D 1983 *The Reflective Practioner: How Professionals Think in Action.* New York: Basic Books

[21] Kolb D A 1984 *Experiential Learning. Experiences as a source of Learning and Development.* Englewood-Cliffs: Prentice-Hall

Part II
Design Development

A Socio-Technical System for the Support of the Management and Control of Engineering Design Projects.

Dr A P Jagodzinski and R Parsons
School of Computing, University of Plymouth, Drake Circus,
Plymouth PL4 8AA, UK

C Burningham , Prof J Evans and Dr F Reid
Department of Psychology, University of Plymouth, Drake Circus,
Plymouth PL4 8AA, UK

Dr P F Culverhouse
School of Electronic, Communication and Electrical Engineering,
University of Plymouth, Drake Circus,
Plymouth PL4 8AA, UK

Abstract

This paper starts by describing the findings of three linked studies of the early stages of the process of electronics design. These studies identify a set of interrelated problems which have been found to exist widely in the industry. The problems lie in the areas of specification, risk management, designers' roles, project planning and organisational knowledge. These problems are compounded by the complex dynamics of project team working which are only rarely amenable to static management and control tactics. Interrelationships between these problems are identified and an integrated approach to their solution is outlined, consisting of a socio-technical system based on the notion of dynamic contingency management, supported by interacting software agents. The paper concludes with short descriptions of the requirements for some of the key components and their roles in the system.

1. Introduction and Background

This work concerns the early stages of electronics design, that is, from the receipt of a specification of customers' requirements to the creation of a detailed functional specification [1]. The aims of this paper are first to describe a set of interrelated problems which have been found to exist widely in studies of the socio-technical systems for electronics design in industry and secondly, to outline an approach to the solution of such problems. Some of the problems have been described separately in previous work, but the importance of the present analysis is to show how problems in each area create different problems in other areas. From this perspective it is possible to envisage a set of interrelated solutions in a linked framework of dynamic contingency management. Our analysis suggests that only such a systemic approach will be capable of tackling the problems identified. The work is based on investigations of industry practice and problems. Three studies in particular have contributed to this paper.

The first entailed a survey of industry practice, CAD tool usage and limitations, design theory and process models. It was based on literature reviews and visits to eighteen UK and European electronics manufacturing firms and eight leading US, Japanese and Korean electronics companies and research institutes. These companies represented the automotive, aerospace, process equipment and consumer products sectors. As far as possible the companies chosen were held (e.g. by Department of Trade & Industry and CAD tool vendors) to exemplify good practice in the industry. This study provided insights into the major problems facing product designers, and enabled a set of best practices to be identified. More details of the study can be found in [2] and [3].

In addition to the general problems, discussed later, the study revealed a number of widely held views which can be seen to characterise the industry. Most of these are well-known, and two in particular set the climate of belief in which electronics design takes place. The first of these views is that time-to-market is the most important priority in the product design life cycle. Inevitably, when this belief is universal it is also true. Among its consequences is the fact that optimal design is not usually a realistic possibility and that satisficing design is a necessity borne of the markets. Secondly, there is widespread recognition of the benefits of concurrent engineering approaches to product development. However, more traditional prototyping approaches, giving a serial pathway for product development, are more commonly used because they are easier, requiring less sophisticated communication systems, project control and management.

The second and third studies have entailed protracted observation of design teams in one of the UK companies, a leading 'silicon foundry', sampled in the first study. The objective of these studies was to flesh out the generic problems which were found in the industry with a detailed understanding of their underlying causes.

The second and third studies both focused on two design teams, working on major projects in one company, in two ways:

(a) an ethnographic study of the social activity within design teams, providing a description of project management, work practices, problems, errors, decision-making events and information pathways during the early phases of design. This part of the study represents an extension, in an industrial setting, of the work reported by [4], on social processes among designers.

(b) a systems analysis examining current usage of computer based support tools for CAD, project planning and control, documentation and designers' individual record keeping.

This work has taken the view that considering design as a social activity is particularly important for the early stages, where 50% of the cost of manufacturing a product may be committed within the first 10% of the design process [5]. This part of the process is concerned with turning what can be quite vague and conflicting aims into hard engineering targets. Due to the increasing complexity and size of artifacts this has through necessity become a team process, which must allow a high degree of flexibility and adaptation in the interpretation and exchange of ideas. For this reason it is not amenable to the tight reductionist modelling which has been applied e.g. [6], to the later stages of design. Concurrent Engineering (CE) '... the concept of running design activities and reflecting the effect of design influences simultaneously.' ([5] p. 31) aims to address this situation by developing and applying multi-skilled teams supported by appropriate technologies [7]. This approach moves from the hierarchical to the 'adhocracal', placing a greater burden on co-ordination, and the communication and storage of information.

The point is that flexibility and adaptability are achieved by what is primarily a social process of debate and interaction between people. This process is concerned with juggling conflicting aims and constraints to produce an acceptable definition of the problems (as outlined in figure 1 below), which are addressed in the later stages of design.

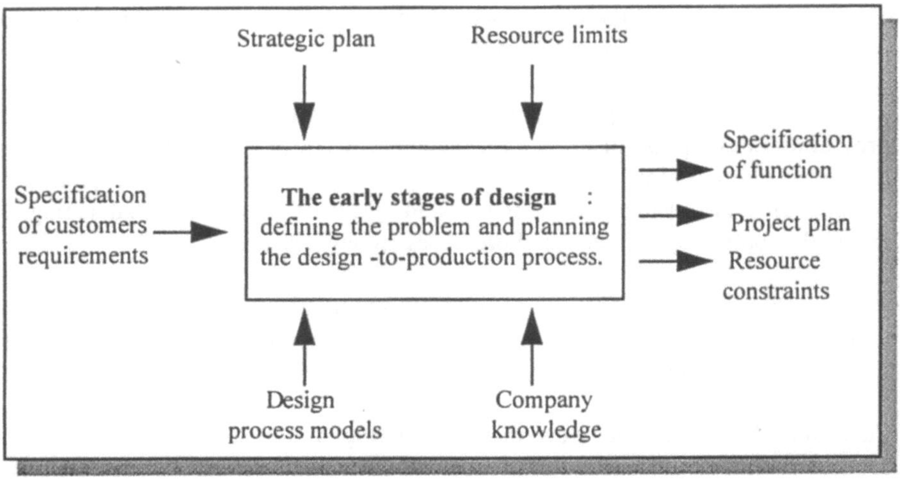

Figure 1. The early stages of electronics design.

2. Methods

The aims of the research were to derive a specification of requirements and specification of function for a system to support the management and coordination of electronic engineering design projects, nicknamed "Surrogate design manager". The existence of social and cognitive psychologists, electronic engineers, systems analysts and HCI practitioners, and generous access to design teams in industry, provided the opportunity for a thorough study of the technical and social aspects of design practice and problems in industrial settings.

However, such a potentially rich study also required the development of a coherent method to enable the insights of the different disciplines and studies to be integrated. When the research started in 1990 existing methods were limited to combinations of Soft Systems Analysis [8], Object Oriented Analysis [9], and User Centred Systems Analysis [10]. Since then methods have emerged which recognise the need for the integration of different approaches, and the method which evolved for the present work can be seen to resemble some of these.

2.1 Requirements Engineering : an Integrating Framework for Research Methods

Many models of the process of systems development exist which describe to varying degrees the development of various types of systems, from simple databases to expert systems. However, [11] argue that whatever the application type the process may be regarded as comprising three essential stages, these being:

- Capture and approval of a satisfactory set of systems requirements.
- Generation of a design that satisfies these requirements.
- Realisation of an implementation that conforms to this design.

(after [11])

The development of a fully functional system was beyond the scope of this present project. Instead, the project has focused upon what may be termed, socio-technical requirements engineering, completing the first stage in the above sequence. Recent trends in systems development have acknowledged a number of key issues which confirm this focus and the project's approach. [12] describe three of these issues.

Firstly, requirements engineering may be regarded as a project in itself with its own life cycle. For example, [13] propose that the requirements life cycle has a three phase cycle of elicitation, expression and validation. Whilst [14], suggest a similar sequence but add that the evolution of the requirements should be accomplished through stakeholders using scenarios to challenge proposed requirements.

Secondly, [15] and [12] criticise concurrent software developments for focusing on the characteristics and structure of the solution rather than the problem. This situation occurs according to [16], because the situation of the requirements is ignored. [16] states that requirements are information and all information is situated, consequently it is the situation that will determine the meaning of the requirements. He proposes that both social and technical factors must be considered, and that this would mean the integration of ethnographic techniques.

Thirdly, [12] state that 'The oldest and perhaps most widely shared piece of conventional wisdom is that requirements constitute a complete statement of what the system will do without referring to how to do it' (Page 17). We would agree with the views of both [12] and [16] that in an open system, requirements are highly unlikely to ever be 'complete' due to the inevitable issues of change and flexibility that must be accounted for. Consequently, this 'incompleteness' must be accepted and managed, through stipulating and communicating the level of completeness that has been achieved.

2.2 Systems Analysis and Development Method.

The philosophy underlying the requirements engineering method which has been employed, combines the strengths of a number of approaches to produce a system analysis method suitable for establishing requirements for supporting electronic design. The source methods include socio-technical, soft systems, human factors and object oriented approaches. In addition to these approaches the work also employed more conventional information systems analysis techniques in the initial stages of the study. Structured techniques such as data flow diagramming were applied to capture the way in which control of the early stages of the design process is maintained and how the knowledge of the design teams is communicated and stored; in terms of both computer and non-computer based methods.

[17] report the successful use of Soft Systems Methodology (SSM) [2], for a study of project management requirements for a medium sized electronics manufacturing company. Their rationale being, as in the present work, that it provided a tool for improving the understanding of a complex and unstructured problem situation. SSM can also be regarded as a tool to help participants learn about their situation, and stimulate the identification of opportunities for change to improve the current environment. SSM does not aim to generate a normative solution for an organisation, it seeks to generate debate of visions of the environment by the organisation. Through the social activity of debate optimal enhancements to support the environment may be identified and agreed. As such debate may be considered a catalyst for change which can assist the development and implementation of a support system. We have utilised this approach as our development process to generate and evaluate the requirements for support in this environment.

The analysis involved studying socio-technical 'primary work groups'. That is, a system which carries out a set of activities in an identifiable and bounded subsystem of a whole organisation involving people and technology [18]. The primary work groups being two of company's design teams, which exist as human activity systems within their design department. We adopted the approach of the socio-technical paradigm, coupled with the essence of SSM as the underlying philosophy for our analysis. By this we mean that our focus is the workgroup, not just the individuals; any technology used should support workers in their tasks; that the environment is a complex social system, where a normative solution is not possible; and we aim to generate requirements for improving the current situation. To guide our system development we have adopted the principles HUSAT's Human Factors Guidelines [10], which aims to ensure that the resultant system is user centred. The overall approach is illustrated in figure 2 below.

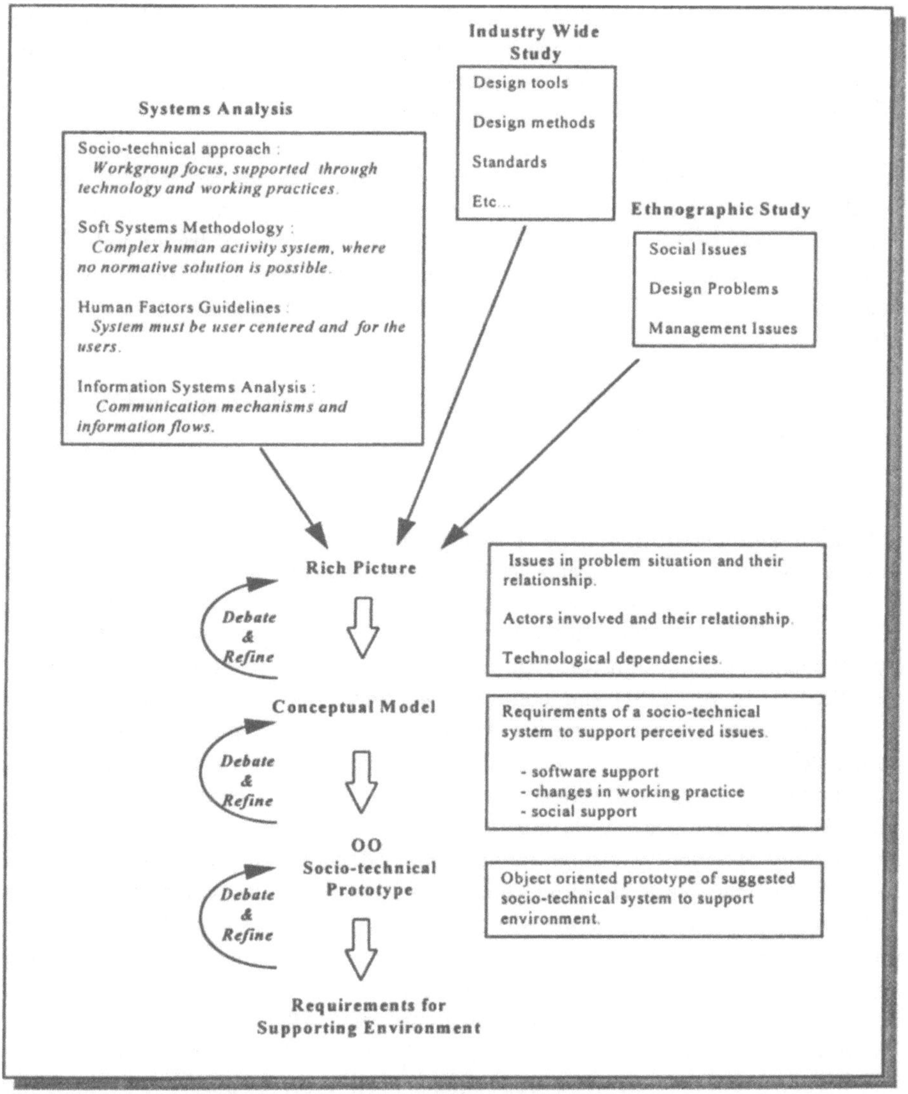

Figure 2. The integration of systems analysis and design techniques

Ignoring day to day articulation work has led to the failure of systems because of a lack of tacit understanding [19] and [20]. [21] illustrate the implications of this concept well by pointing out that during industrial action workers will 'work to rule'. That is, that they will only perform tasks as described and directed by their official rules of work. The approach of collaboration between software engineers and sociologists has been used with some success in various projects at Lancaster's CSCW research centre [22]. [21] report that although this approach proved successful, they found difficulty in extracting the systems requirements from the ethnographic data. Indeed, this has been the main problem encountered with utilising ethnography in systems development. [23] and [24] argue that traditional structured analysis and development methods ignore the important element of the systems functioning within an environment. Hence the adoption of SSM and the socio-technical paradigm. The integration of the longitudinal ethnographic study and systems analysis method is discussed below.

As the above diagram illustrates, a crucial element of the projects work is the way in which these requirements were evaluated as they evolved. The deliverables from each stage of the research are less tangible than those of traditional systems analysis, and consequently require appropriate techniques to evaluate them. To accomplish this cognitive walkthroughs and thought experiments have been utilised.

2.2.1 Validation and Verification - Using Cognitive Walkthroughs and Thought Experiments.

At each stage of the requirements development we evaluate on two issues, namely validation and verification. [25] neatly defines validation as being 'are we building the right product ?', and verification as 'are we building the product right ?'. The target system was for the vast majority of users, intangible and hard to visualise. Consequently a suitable method for evaluating what was effectively hypothetical system design was required. This is the aim of 'thought experiments' (see [26], [27], & [28]) and cognitive walkthroughs (see [29] & [30]). Both thought experiments and cognitive walkthroughs utilise expert panels to reason about, debate and appraise a given theory, before conducting a fully functional experiment. This approach reflects the essence of SSM, as discussed above.

Thought experiments are fairly unstructured and aim to test the general approach and viability of a theory, through domain expert debates. Hence this approach was suitable for the earlier, more theoretical stages. In contrast, cognitive walkthroughs (CW) are more structured, aiming to test a theory through predetermined goals. [29] and [30] cite improvements in evaluating systems when applying CW. This approach requires a more detailed definition of the artefact to be evaluated, so that it will be used upon completion of the prototype system.

The development of the prototype system may be regarded as a method of *buying* information about the perceived requirements [31]. It serves purely to evaluate the requirements and their viability. The prototype will provide the users with a usable and recognisable mapping of a case history of one of the design teams studied, on to the support system proposed. This approach permits us to illustrate a tangible scenario that the stakeholders can readily challenge. This is a method of evaluating requirements supported by [14].

The preceding sections have discussed the philosophy and method underlying the research. The following section describes the project's research cycle in terms of the socio-technical systems development.

2.2.2 The Research Process : Evolving Requirements

The project is fundamentally a bottom-up investigation, in that there was no preconceived solution type to be applied. Instead the project was given the task of identifying any issues which were detracting from the effectiveness of managing the design process. The research entailed cycles of elicitation, expression and evaluation to determine the requirements for enhancing the current socio-technical support system. This process is illustrated in figure 3 below.

Initially, an elicitation stage was completed using the methods discussed in the previous section. In addition, contemporary systems analysis techniques, such as data flow diagrams, were applied to capture the way in which control of the early stages of the design process was maintained and how the knowledge of the design teams was communicated and stored; in terms of both computer and non-computer based methods. The analysis was carried out over a two month period, focusing upon the two design teams being studied by the psychology based project. The work involved analysis of the company's existing documentation, both hard and soft. This was followed by a series of interviews, both structured and unstructured, with staff in the design teams, design support team, and Information Technology Services.

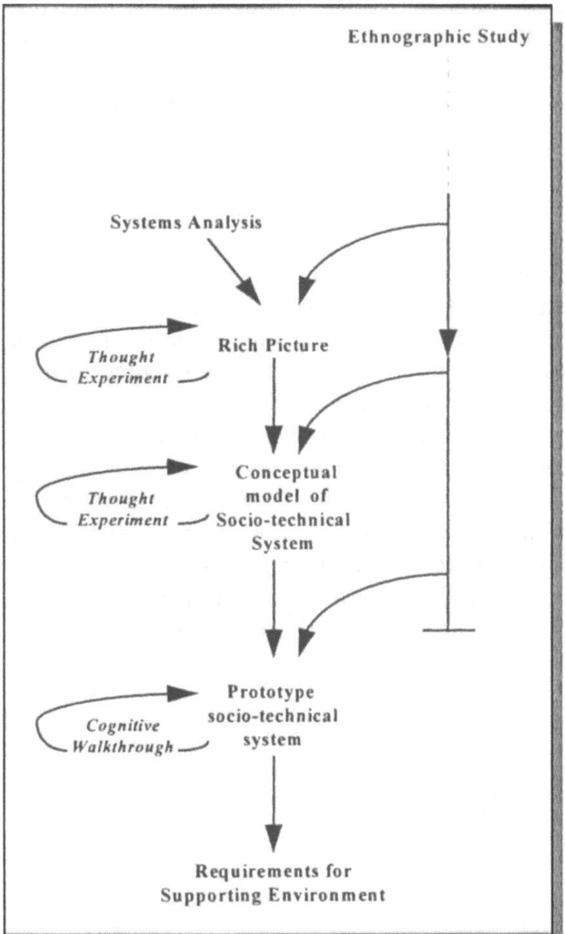

Figure 3. The research process.

This is the RE cycle as proposed by Jarke et al (1994). However we would use the term evaluation rather than validation, for the reasons explained in section 2.2.1

In addition, vital design team activities such as project review meetings and process mapping exercises were attended. This stage resulted in the generation of a rich picture of the current environment. This was evaluated and refined through thought experiments, as discussed above, in conjunction with company stakeholders.

Following the initial elicitation stage, a conceptual model which addressed the perceived issues was developed. Again this was evaluated and refined via thought experiments in collaboration with the company. From the results of the second stage of the process, an object oriented prototype socio-technical system is being developed. This will be evaluated via cognitive walkthroughs by the company. Whilst the final conclusions of the longitudinal psychological study are still undergoing completion, as illustrated above, the ethnographic data has provided vital insights to ensure that the specified requirements reflect the social needs of the environment.

2.3 Industry-wide Study

This essentially systems analytic study was carried out between 1990 and 1991. Data were obtained by means of structured interviews which defined the categories in which information was sought but left the respondents to decide the form and content of the replies. Categories were chosen to elicit an understanding of how each of the twenty-six companies currently developed its products. Interviews were conducted by two members of the research team who, together, questioned up to four design and production managers per visit. In two cases Managing Directors were also questioned. In each visit interviews were spread over up to two days, each lasting from three to six hours. Very senior design, R and D and executive staff managers were interviewed. In addition, visits included guided tours of design and production facilities and demonstrations of design tools. As far as possible attention was focused on identifying problem areas, although examples of good practice were also sought.

In summary, information was collected in the following areas:

- Nature of design tools and problems with their use.
- Integration of design function with other computer aided parts of the organisation.
- Design environment/company culture.
- Design methods.
- Risk assessment.
- Simulation techniques/software used.
- Engineering change control policies.
- Design/production interfaces.
- Component policies.
- Standards, ie ISO9000 and environmental impact of manufacturing techniques.
- Impact of manufacturing approaches (e.g. JIT, OPT, MRPII) on the design function.

More details of this study can be found in [2] and [3].

2.4 Single Company Ethnographic Study

Two design teams were studied for a period of 40 weeks during the early stages of the design process. That is, from customers' statements of requirements to the specification of designs at a functional level. One team was designing a new RISC processor, and this was considered to be an innovative project with high levels of uncertainty both in the technologies to be exploited and the software tools to be used in the design process.

The second team was designing a variation of an existing chip. This product was intended to be sold to consumer product manufacturing companies to be customised by them for their own requirements. The design project entailed extending the chip's functionality.

In this way the research was able to compare and contrast practices and problems at opposite ends of the continuum of new knowledge versus existing knowledge, which prove to have significant implications for management and control.

Analysis of detailed ethnographic data is still in progress. However, significant elements of the studies have been described and together with the systems analysis described in sections 2.2 and 2.3, form the basis for the remainder of this paper. These elements include a company familiarisation study, an analysis contrasting the views of the project leaders with those of the company's management, and a technical report based on an extended series of interviews with project leaders (PL). Excluded from this paper are the conclusions of a detailed ethnographic analysis of diary studies.

The PL interviews study is of particular interest and the following points are relevant here:

a) multimethod qualitative approach - the interviews form part of a structured longitudinal ethnographic study combining the following qualitative data gathering techniques:

• direct observation and shadowing - here the investigator produced a narrative record of observations carried out on normal working activities of target personnel in selected domains of the company.

• depth interviews - semi-structured depth interviews were carried out with preselected personnel to identify goals, working practices, interpretative schemes, judgmental and attitudinal responses, etc. The PL interviews were all carried out at roughly the same time at the start of the study to identify common themes and issues at that level of management within the company.

• retrospective probe interviews - because of their intrusiveness, conventional diary keeping methods were judged inappropriate in this context, and longitudinal data were gathered using weekly retrospective probe interviews with selected personnel in target product teams. These interviews followed a fixed protocol, and gathered attitudinal and judgmental data throughout the history of these teams.

b) Data sampling - this was a key consideration, and carefully planned. Three levels of sampling were identified :

• design management focus - the focus of the interviews and subsequent longitudinal studies was on the design and management activities of PLs. The rationale for this was established in the company familiarisation study, which involved carrying out a contrastive analysis of senior managers' and project leaders' interpretations of the management role. This analysis found that the company's senior managers did not have a high level of interaction with product development groups, and that PLs fulfilled the main operational design management functions in five areas: technical leadership and decision making, group integration and motivation, staff development and appraisal functions, product team advocate, reporting to and interfacing with senior managers and other departments, customer interaction and product specification. In other words, it was the PLs not the managers who were responsible for the day-to-day management of the product development process, and it is their activity that the target system should seek to simulate and/or support.

• product design focus - of the five product groups surveyed, two working on new microprocessor products were selected for their contrastive potential; a group with a reasonably well-developed product specification and moderate levels of uncertainty, and a group working with a very well-defined specification and lower levels of uncertainty. The purpose of this contrast was to detect differences in design process attributable to differences in innovation and complexity (i.e. cells 2 and 4 in the [50] model, see section 4).

• personnel focus -in addition to the PLs of the target product groups, one other central group member was included in regular data sampling. The intention behind this was to provide a reliability check on the identification of critical occurrences in the design process, and to allow triangulation on event interpretation and characterisation.

• Issue salience - a major justification for the use of structured interviews in this setting was the criterion of issue salience. This criterion is based on basic cognitive science theory, and assumes that respondents sample their memories for instances of occurrences triggered by probe questions, and retrieve vivid, recent, personally significant, or otherwise more memorable, instances of these occurrences. In other words, respondents are assumed to produce reports on the, most salient occurrences, rather than an exhaustive and unbiased recall of events. The present study capitalises on this process: we chose to be guided by respondents' intuitive judgements of what is relevant and significant, rather than our own presuppositions. A further point worth emphasising is that the company's engineers generally had little patience with fashionable management ideas or time for 'navel gazing': to the extent that they volunteered comments on design management problems and practices, we concluded they must be of some magnitude and significance for them to draw their attention.

• Interview Procedure - 15 PLs and their 4 managers participated in semi-structured depth interviews, lasting an average of between one hour and an hour and a half. The managers were interviewed in order to provide a context for the PL interviews. These interviews have been not analysed in detail. The PL interviews probed PLs for their work role, what it encompassed, and with whom they interacted. The interviews also covered the areas of work style, communications, project control, design knowledge, the design process and relationships at work. The PLs' occupational background and experience were also discussed and any other area the subject wished to cover. It was emphasised from the start that individuals identity would be protected, with the research interest one of discussing common themes rather than individual differences.

3. Summary of Major Findings : Key Problems in the Design Process

The three studies, as one would expect, produced qualitatively different findings. The industry-wide study, which was conducted essentially as an exercise in systems analysis, and the single company systems analysis both identified finite problems and issues. Some of these are well known and widely recognised, although not usually clearly articulated. Furthermore, they were generally seen as separate problems rather than critically interdependent ones. These problems were also confirmed by the direct observation and depth interview work.

The ethnographic study, at this point based mainly on project leader interviews, naturally emphasised management and control issues. These were distributed across a very wide range of activities only some of which coincided directly with those identified in the systems analytic studies. However, the study provided valuable evidence of generic problems arising from the increasing technical complexity of design and the need for flexible and dynamic tactics for project management and control. To this extent they supported the findings of the systems analytic studies regarding the complex, dynamic interdependence of the problems in the design process.

3.1 Findings of the Systems Analytic and Observational Studies

Six major problem areas emerged from the studies, in most cases in an entirely separate and fragmented fashion. This, as [21] discuss, is a typical outcome of ethnography when applied in this way. We present the findings here in a way which seeks to show that some of the problems are closely interlinked, sometimes causally. All of them can be seen to have major impact on time-to-market.

3.1.1 Specification Intent

The first stage of the engineering design process is, in most models (e.g. [1]), normally some sort of feasibility study or market needs assessment which aims to match some identified customer need with the company's design and manufacturing capability. The result of this stage should be a specification, which, typically, may lie somewhere on a continuum ranging from:

> -*hard* : a statement of detailed functional requirements which the product should satisfy, with budget and time scales for design testing, production and marketing.

to:

> - *soft:* a vague wish list of areas in which potential customers have expressed an interest and for which new ideas are sought, without specific time scales or budgets.

The specification usually provides the design team with its starting point. Both forms of specification may be equally valid, as may an intermediate form, depending on the conditions in a particular market. Ideally the soft form would be expected to develop into the hard form before detailed design took place.

However, this is where difficulties occur. It would seem that designers can be unclear as to the intent of the specification with possible consequences as follows:-

a) specifications sometimes lack information about the intent of the originators so that it is not made explicit whether they are hard, soft, intermediate or a mixture of all three;

b) specifications pass down the hierarchy from design manager to project leader to design engineer. In this journey soft specifications sometimes become hardened up as each level adds its interpretation to the original request for ideas. Design engineers thus receive the impression that the specification is hard when it still should be regarded as soft. Alternatively, the hardening is intended, but the difference of intent is not evident.

c) thus for soft specifications, depth first, detailed design may take place when breadth first, outline consideration of a wide range of possibilities would be more appropriate. Inevitably, much of the detailed design work is wasted when an alternative option is adopted;

d) when a soft specification is being developed it may not be circulated to the right people, for example for reasons of commercial confidentiality, so that when design starts, changes to the specification are needed to accommodate additional knowledge;

e) communication channels do not always exist to allow discussions between designers and customers to refine details of technical requirements;

f) for soft specifications engineers can, for various reasons, be reluctant to express their views in writing. Informal discussion of issues would elicit a balanced view, but the right forum does not always exist.

The problems of specification can be seen to be closely related to the problems of Organisational Knowledge (3.4), Management of Risk (3.2) and Project Planning (3.3).

3.1.2 Management of Risk

Risk is an inevitable concomitant of innovation, and can be seen to be directly proportional to it. However, the degree of innovation actually required by a particular design may vary considerably from case to case. [32], [33] and [34] all identify at least three degrees of innovation by which a particular design can be characterised, for example "routine design", "innovative design" and "creative design" [32]. This distinction is often not made at the start of a design project sometimes because the intent of the specification is not clear. The consequence is that levels of innovation in the design may vary throughout the project in an uncontrolled, *ad hoc* fashion. For example, a new technology or process may be incorporated into the design part-way through the project, bringing significant delays with it.

The net effect of such *ad hoc* changes in the level of innovation may be beneficial (if it results in a better product), or harmful (if it delays the release of the product in the market), or even both. However, what is important here is that such effects often seem to happen in an uncontrolled way. The decision on levels of innovation could be made deliberately on the basis of a conscious assessment of the associated risks, but this was often seen not to be taking place.

3.1.3 Project Planning

The industry is generally well acquainted with all of the classical project planning tools and techniques such as activity networks, critical path analysis, PERT, Gantt charts and so on, and widespread use is made of computer based tools in this area. Nevertheless, project planning may be regarded as difficult and not always successful. Sometimes "planning" is a matter of fitting the chart to match actual events. Engineers may give themselves extra time in their estimates "just in case something goes wrong". Sometimes time scales are not very clear or have not been decided upon because of unknowns or unknown customer requirements. Alternatively, engineers' time estimates are described as "notoriously optimistic".

There are several reasons underlying this problem. First, if design is innovative then it is, by definition, going into the unknown so that time estimates are likely to be much more vague than they would be for a well-defined task such as maintenance.

Secondly, the tools being used are not always perceived to be appropriate to projects with a high degree of uncertainty.

Thirdly, if planning lacks credibility then the possibility arises for a snowballing effect. For example, if a particular engineer is not given a hard deadline for a particular piece of work then it is easier to divert him to some other task, so that more slippage occurs on the first project, and so on.

More fundamentally, problems of project planning can be seen to arise partly from lack of attention to risk-management (3.2) which means that neither a clear design path nor an acceptable level of innovation is set at the feasibility study stage. This problem is compounded by specification (3.1) when insufficient guidance is given as to where on the possible continuum of specifications, from hard to soft, this one falls, and thus what sort of approach is required of the designers. Without a clear indication of what is expected in these areas, it seems likely that planning will sometimes be, to some extent, mere window dressing.

3.1.4 Organisational Knowledge

The knowledge that a design project can draw upon exists in two complementary forms, namely human memory and externally stored knowledge, each of which will be considered in turn.

a) Transactive memory

The role of the individual designer's memory although obviously implicit in their work has not been the focus of the present study.

However, what did emerge in several ways in the studies was the need for knowledge about what other people know. For example, in the specification process it was important for the initiators of the project to know who should be consulted in order to cover all of the important aspects of the proposed design. As a second example, it was fundamental that members of a design team should know who in the team, and outside of the team, they should consult about particular issues, such as the characteristics of different technologies. An important component of the role of Project Leader was knowledge of the specialisms within the team, and outside it in the organisation at large. This type of knowledge about who knows what, has been termed by [35] as "transactive memory":

"*A transactive memory system is a set of individual memory systems in combination with the communication that takes place*".

(p186)

"... *one person has access to information in another's memory by virtue of knowing that the other person is a location for an item with a certain label. This allows both people to depend on communication with each other for the enhancement of their personal memory stores. At the same time, however, this interdependence produces a knowledge-holding system that is larger and more complex than either of the individuals' own memory systems.*"

(p189)

Transactive memory was clearly implicit in all of the design activity examined in the study. It emerged explicitly (not of course by that name) only when it was perceived to be failing to work properly.

In ideal conditions, such as geographically co-located design teams of less than 10 members, there seemed to be no problem. However, in other circumstances such as during specification, problems arose from the need to consult more widely and over long distances. This difficulty is also described in section 3.1, and can be seen to arise from the absence of a comprehensive record of who knows what.

The investigators were left with the uneasy suspicion that there may be more problems which arise from transactive memory failures, particularly in large organisations. Furthermore, the limitations of the normal communication mechanisms of transactive memory may prevent organisations from even being aware that some of these problems exist.

b) Externally stored knowledge

Although there are many types of externally stored knowledge such as parts catalogues, standards, procedures manuals and so on, the key issue facing engineering design companies in this area is the capture, control and recall of product design information from design and manufacturing staff. Engineers almost universally dislike doing documentation because they feel it interferes with design (See also section 3.5, Roles in design teams). A design's documentation will normally hold the minimum amount of information necessary to specify its form to the next step of the design and manufacturing process. This economy of notation is supported by CAD tools which only store information that is explicitly required to achieve the efficient development of a new product [36].

Our studies have shown that once a design's documentation has been archived it is unlikely to be used again. This is likely to be because the information stored in such databases, paper or computer based, is so terse that it is meaningless to any but the most determined and well informed enquirer and, secondly, that usable access mechanisms do not exist.

In order to be accessible and useful, such design knowledge would have to include much additional information such as:

- functionality
- the reasons behind particular design decisions
- interfaces
- design validation criteria
- test strategy
- costs
- identifiable risks
- standards adhered to
- manufacturing issues,

and so on.

The production of such documentation could entail additional effort and could also be unpopular with design engineers (see section 3.1.5, Roles in design teams).

The problem arising from this situation is that externally stored knowledge about previous designs is effectively unavailable even though such designs might easily be, at least partially, reusable. Clearly, reuse of previous designs could greatly reduce the risks and time scales of new design projects. Similarly, knowledge of previous, unsuccessful designs could help project teams to avoid exploring fruitless options.

3.1.5 Roles in Design Teams

We observed situations in which there seemed to be clear conflicts between the demands of the design task and the designer's perception of his or her own role. In most cases engineers saw their prime function as designing, implementing and testing electronic artifacts. They judged themselves, and felt judged by others on their performance in these tasks.

On the other hand documentation of the design can also be seen to be very important, as explained in section 3.4.2, and yet typically designers regard the activity of documentation as detracting from their performance in their principal role so that they may do it badly. Some anecdotal evidence exists for the creation of more or less formal additional roles in design teams with titles such as "Technical Author" and "Librarian", suggesting that recognition of such roles may be useful.

Another area in which there was role conflict was in cases where experienced designers had been promoted to management positions. This could cause several problems. First, their technical expertise was still in demand (because of the transactive memory system, section 3.1.4) so that they were diverted from their management task and yet were not being properly recognised in their secondary role. This situation could be alleviated by the creation of the title of Technical Consultant, or something similar, which would enable the dual nature of the role to be formalised.

Secondly, when technical experts were also managers it was possible for their objectives to be in conflict. For example, adherence to a management plan might result in a risk-free reuse of an existing design whereas technical creativity and excitement may pull towards a riskier, innovative solution.

Clearer definition of role would alleviate such tensions by at least enabling the dilemmas to be recognised and arbitrated consciously.

3.1.6 Company "Culture"

Impressionistically, clear differences could be perceived between companies on issues such as their commercial history, market sector norms and priorities, the extent to which they embrace change and so on. However, we have not addressed these issues even though in some cases they might cause problems. This is first because they lie outside of the scope of our studies and secondly, because they tend to be sectoral or particular rather than generic.

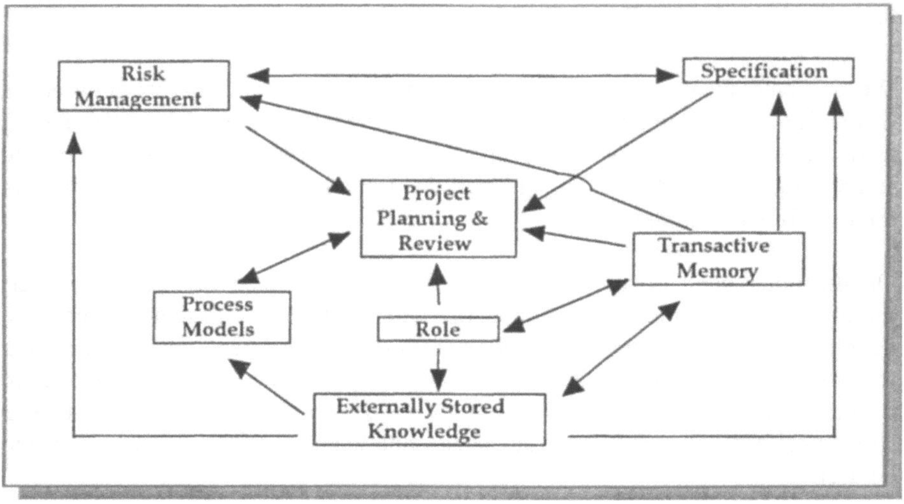

Figure 4. Observed problem areas in engineering design projects and the relationship between them.

3.2 Findings of the Ethnographic Studies to Date

3.2.1 The Nature of Management and Control

The main body of ethnographic evidence at this stage of the research derives from the project leader (PL) interviews. What emerges from this evidence is the existence of a small set of management dilemmas which confront design managers. These management dilemmas present themselves to the PLs as real problems which they have to resolve, but unlike technical difficulties that respond to creative design solutions, these problems are recurrent and sometimes intractable, and have to be addressed repeatedly throughout the product development process. In other words, they are problems that have to be managed rather than solved, and are probably best thought of as system states which have to be maintained within acceptable bounds rather than held to fixed values or alienated altogether. Good design managers are therefore project leaders who are capable of steering the design team on a satisfactory trajectory within these bounds, rather than simply passively fitting management solutions to eventualities as they arise. Good managers are capable of reading the, situation, and responding to perturbations that threaten to push the team outside acceptable bounds with corrective measures. The task facing the project leader is therefore to maintain a complex and dynamic equilibrium over the course of the project.

What the interviews show is that managing design is a process involving responses to design activities and outcomes on a continuous and dynamic basis, inputting corrective measures where necessary to keep the design team on target. Tactical shifts in management structure are actually this process, the real design management issues facing project leaders. To understand this process, we need to be able to identify relevant contingency factors (i.e. occurrences which call for corrective management action), as well as good management practices (i.e. appropriate, corrective responses). In line with our current thinking, the appropriate metaphor here is *situated cognition* [37], the complex process of fitting responses to patterns of indicators, rather than the production rule metaphor.

A dynamic, contingency approach to project management is discussed in section 4.

3.2.2 Management Dilemmas

Two principle management dilemmas emerge from the interviews: the management of *workflow interdependence*, and the management of *design team integration.* These represent recurrent and refractory demands on PLs' management skills, and appear to be distinct concerns underlying the "problem frequency" axis [50], see figure 7. How PLs respond to these dilemmas has implications for the 5 problem areas identified in section 3.1.

a) Workflow interdependence

Workflow interdependence deals with the consequences of the coordination requirements involved in complex design problems where the activities of the team need to be coordinated to result in an integrated product design. This area corresponds to the characterisation of design management by [50], as managing information flows through successive stages of the design process. What the interviews show is that this is a central and recurrent dilemma for the PLs, *but it is not the only dilemma*: other concerns orthogonal to workflow interdependence emerge in this data.

The PLs recognise that workflow interdependence is a scalar, and that different responses are appropriate to different task situations. They represent this distinction as a state of tension between alternative modes of operation that requires active management. This tension is reflected in the following:

- variation in how closely PLs monitored the work of engineers in the team according to task demands and engineering skill profile. This variation could either reflect differences between PLs in their current management concerns, or stylistic preferences to manage the team in particular ways.

- a fear that the company employs inadequate and insufficiently formal procedures for project decomposition, planning, and scheduling, leading to an inappropriate reliance on judgement and ad hoc adjustments later in the design process. Thus concern recognises the value of an ad hoc approach to management, but only if combined systematically with a planned and scheduled approach.

- a recognition that time scale slippage is inevitable (leading to the accepted usage of slip charts alongside milestone charts), that the design process must be capable of responding to this eventuality (in the face of pressures to mechanistic time scale adherence from senior managers).

- a recognition of the. value of project planning tools under complex design conditions, where multiple subassemblies are involved. This reflects an appreciation of the exponential increase in workflow coordination problems as team size or problem complexity increase.

- a recognition of the importance of fully open communication and feedback channels between engineers when mutual adjustment and reciprocal interdependence is the priority concern. This is also reflected in comments about the physical proximity and open plan arrangement of the teams' working environment: that such an arrangement is essential to workflow interdependence, but that it is assumed to take place, and that without astute management and regular monitoring this may not actually occur. In other words, the open plan office is also considered to have a potential for facilitating information loss since ".. things fall down because there's an assumption that people hear things and know what is going on".

- a recognition that inappropriate workflow arrangements are often accepted by default when communication difficulties within the company impede mutual adjustment. Under these circumstances, a sequential workflow arrangement (where everybody waits on hold whilst a critical problem is solved) may be reluctantly accepted.

b) Design team integration

Design team integration concern focuses on the difficulties PLs faced in maintaining their teams as functioning units. The conflict here is a well-known and much studied problem of establishing and maintaining equilibrium within the team between two opposing motivations, the motivation to pursue and accomplish task objectives (the task motive), and the need to attend to personal concerns and relationships within the team (the socio-emotional or group integration motive). Ever since the classic work of [38], this balancing act has been known to be a primary concern of all team managers. The reason for this lies in an obvious paradox: whilst a team that spends too much socialising may fail to address its primary task, a team of people who cannot get on with each other are unlikely to be able to cooperative effectively. In unstructured groups, these concerns are often dealt with by different individuals (known as the informal task and socio-emotional leaders), who act in ways that compensate and balance these competing motivations. In these groups it is known that one individual soon emerges as the task leader, actively pursuing task goals and drawing the group's attention to these. This person is not particularly well liked, and tends to build up resentment and irritation amongst other team members. At some point, another leader role emerges, in which a second person acts, through humour, friendly remarks, expressions of group solidarity, etc., to defuse this tension and allow the group to settle back into the task. (It is worth noting that author [39]'s group roles are really no more than a more detailed characterization of these two basic functions.)

In an organised design group, both concerns are likely to fall into the PL's lap. It is possible that, given time, other members of the design team can assume some aspects of these two leadership roles. But design teams are often short-lived, assembled on the basis of expertise not personalities. In these circumstances these competing concerns will need be managed by the PL. Such concerns are reflected in the following comments arising from the interviews:

- the experience of conflict between PLs description of themselves as technical experts or as managers of the design process. This conflict is often acute, and PLs differ in their response to it, some preferring to view themselves as technical consultants, others as full-blown managers.

- a consequence of this conflict is that some PLs experience role overload: not only had they task functions to fulfil as team members, they also had to deal with management problems. This was expressed not so much as a dislike of having "managerial" functions but simply not having enough time to do a run-time design role and also manage. This was expressed in, "...you have a workload yourself which needs to be done and you can't always devote the time necessary to sit down with them and ask them how it is going". Also expressed as, "...ideally you would want, if you were a project leader, the main task to be the project and you wouldn't want to have a large technical load. ...in reality you never have enough people with enough experience for some of the tougher problems".

- the recognition that demonstrating their technical expertise is a criterion for being accepted as a manager, where reputations are made or broken on the basis of visible technical competence. This demonstrates a critical interdependency between the two leadership roles when they are embodied in a single individual. A further reflection of this was the lack of formal authority perceived to accompany the role, "they think I have got some sort of power, but I haven't really - more of a paper tiger than anything".

- a recognition that managing personal issues with the design team is often decisive for team morale, and that newly promoted PLs often neglect these. Some PL are very technically oriented and can neglect some of the morale issues. This is particularly the case with less experienced PLs, i.e. "...you'll find that new PLs tend to neglect all the personal issues and you find that people get disgruntled and things don't get done and people get very demotivated".

- a range of opinions concerning the value of formal design review meetings. Some PLs recognised their task-oriented function, others their political and motivational function. Differences between PLs in this probably reflect stylistic preferences for managing competing concerns.

Mapping these management dilemmas on to the five problem areas specified in section 3.1 is not straightforward. Both have a more or less direct bearing on project planning and design team roles (especially the PL role). There is also some relevance for the utilisation of organisational knowledge. However, at a more general level all of the studies show a concern for the difficulties of managing and controlling complex, dynamic and interdependent problems.

4. Contingency Management

Although the industry is generally aware of the issues that we have discussed in section 3, there is no widely recognised framework for their resolution. A major reason why the problems have arisen is that existing methods for design cannot comprehend the increasing complexity which is enabled by technological revolutions. This problem is getting worse. For example, the company's next generation microchip will comprise 1 billion transistors. This chip is estimated to require 1,800 man-years of design effort using existing tools and methods. Clearly, new tools and methods are needed. The problem is compounded if concurrent engineering approaches are adopted, where even greater coordination skills are required. This section present a rationale for a framework which supports the management of the team process of design. We argue that contingency theory may be used to guide the development of suitable design environments. This is achieved by enabling the adoption of the most appropriate coordinating mechanisms dependent upon the current situation of the design project.

A common assumption when the term 'management' is used is that it refers to control. The environment which we seek to support is the early, conceptual design stage, which is predominantly team based. What is evident from our work is that planning and coordinating the team process are key aspects of the management function. [40] argues that emphasis upon control can suggest a confining or coercive objective. He proposes that control and coordination may be viewed as being the same in principle, where the essence of control is "*on directivity and integration of effort, [and the] required accomplishment of an end*" (page 528). This idea captures our intentions and the essence of the environment that we aim to support. This emphasis upon coordination, rather than control is a view shared by [41], who characterises the following five mechanisms which organisations utilise, to varying levels, to achieve coordination.

a) Mutual Adjustment: The coordination of work through formal & informal communication;

b) Direct Supervision: Individual(s) take responsibility for the work of others;

c) Standardisation of work processes: Contents of the work are specified or programmed;

d) Standardisation of outputs: Results of work can be clearly specified;

e) Standardisation of Skills: Knowledge and training required for process is known.

[41] describes how the reliance upon each particular coordinating mechanism will change as an organisation grows. For example, in a small company of one or two employees the use of mutual adjustment is the most convenient and satisfactory mechanism for coordination. However, if the company were to expand, it would find it increasingly difficult to maintain coordination of a larger number of staff if relying purely upon mutual adjustment. Consequently the emphasis for coordination would shift to other mechanisms, for example direct supervision. [41] represents this change in the following diagram (figure 5).

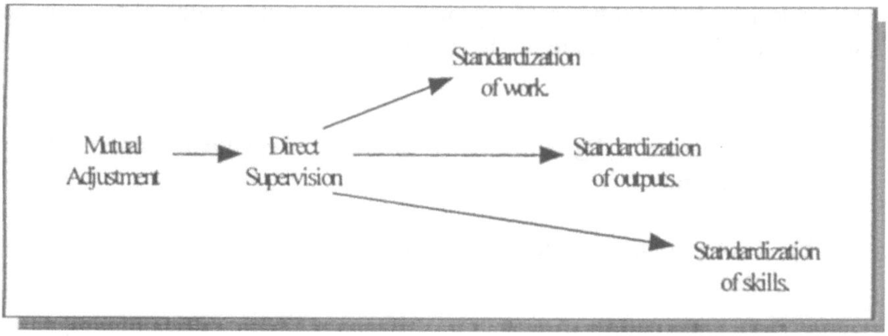

Figure 5. The shift in emphasis of coordinating mechanisms (after [41]).

[41] is directed at a macro view of organizational structure. We suggest that this change of emphasis and reliance upon increasingly formal coordination

mechanisms, may also be seen to take place at a micro level. That is, at a project level within an organisation. Given that an organisation's product and process experience normally evolve, for a given product, through a sequence of projects, it seems likely that the emphasis on each coordination mechanism would change as the knowledge of a product and its related design process increases, as illustrated below in figure 6. In other words, in the case of a design project, its position on the continuum of possibilities for its management and coordination is likely to be governed by the degree of new knowledge involved.

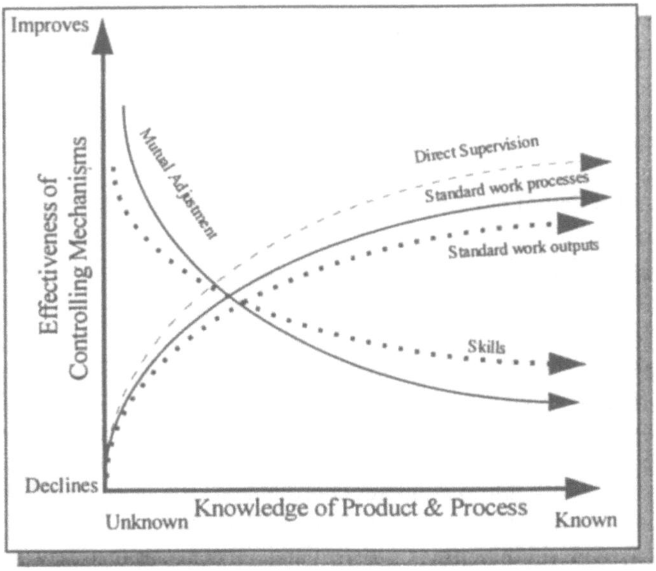

Figure 6. An approximate guide to the change in coordinating mechanism emphasis as a product evolves.

[41] characterises different types of organizational structure, which align with their means of achieving coordination. The following lists the key characteristics of three of these structures :

a) Adhocracy: Dynamic environment, dependent upon mutual adjustment.

b) Professional Bureaucracy: Relatively stable environment, relies upon standardisation of skills.

c) Machine Bureaucracy: Well formalised, reliant upon standardized work processes & output.

The idea of an organisation's structure being suited to its environment has been discussed by many authors (see [42], [20], [43]). Whilst the terminology may vary, for example a machine bureaucracy may be described as a 'mechanistic' structure, the essence remains the same. That is, that dynamic environments function best within an adhocracy or 'organic' structure, whilst static environments are more suited to a machine bureaucracy. These types of structure generally evolve through time as an organisation move from adhocracy to mechanistic. They can be mapped onto the continuum shown above in figure 1. Whilst the above authors discuss these structures at an organisational level their characteristics may also be mapped onto project team structures.

We believe that the essence of these shifts of organizational structures can also be seen in the shift in emphasis of the coordinating mechanisms of design projects, as an organisation's knowledge of a product and processes grows through a sequence of projects. For example, when a product is new to an organisation the project which initially develops the product could be categorised as an 'innovative project', where due to the high level of unknown information mutual adjustment would prevail as the dominant mechanism. In later projects, to modify existing functionality of a product for instance, a project might be described as 'variant', in which the level of knowledge has increased and the mechanisms for achieving coordination have shifted as illustrated in figure 6 above. These mechanisms can be seen to be similar in nature to those identified by [41]. If we were able to characterise projects in such a manner we would be better able to define and support the most appropriate environment for a given project.

Various factors can provide a guide to the most appropriate structure to adopt. Knowing which structure is appropriate for a given organisational environment can be accomplished by applying contingency theory.

4.1 Contingency Theory

Contingency theory is a classical organisational viewpoint, which [20] characterise as stressing that "it is management's responsibility to obtain a *good fit* between the tasks, the environment in which the tasks will be performed and the style of management". It has been discussed by many authors, for example [42], [44], and [45]. [41] proposes that an organisation can decide upon the appropriate structure for their business based upon four 'situational or contingency factors'. This constitutes his 'contingency or situational theory', the four factors being :

For the remainder of this work we shall adopt the term mechanistic rather than machine, as we feel this better describes the characteristics of the environment.

i. The age and size of the organisation

ii. The organisation's technical system (ie. whether it is regulating, sophisticated, automatic etc.)

iii. The organisational environment (ie. stability, complexity, diversity, hostility etc.)

iv. Organisational power relationships (ie. control which is external to the organisation, personal needs of staff, fashions etc.)

Whilst the majority of studies have focused upon applying this model at the organisational level, research has also been conducted at a work group level. Both [46] and [47] applied contingency management at the work group level, within a police force and information systems development company respectively. Whilst the work of [47] and [46] proved to be inconclusive, both regard contingency theory as an appropriate model to adopt at the micro level of work groups. In addition, studies by [48] and [49] concluded that the most effective means of communication and coordination were affected by the level of task uncertainty and workflow interdependence, in accordance with the model of coordinating mechanisms presented above. A common conclusion of research at the micro level has been that contingency factors are difficult to make generic for use in other domains [46]. All of the studies discussed above have taken place in different domains, from systems development to police forces, and whilst their findings are of relevance to our level of interest they did not focus upon an environment similar to design.

More pertinent work has been conducted by [50] in the mechanical engineering domain. [50] present a contingency model for engineering design management. They state that designs fall within one of four cells. These being creative, intensive, incremental and complex design. They suggest that management structures should range on a continuum from organic (ie. an Adhocracy), to mechanistic (ie. a Machine Bureaucracy). The most appropriate structure for an intensive design is organic, whereas an incremental design would require a mechanistic structure. They have identified four contingency factors for assessing the position of the design within their cell structure. The factors being problem structure, amount of information, type of information and problem frequency. Figure 7, below shows the relationship of these ideas.

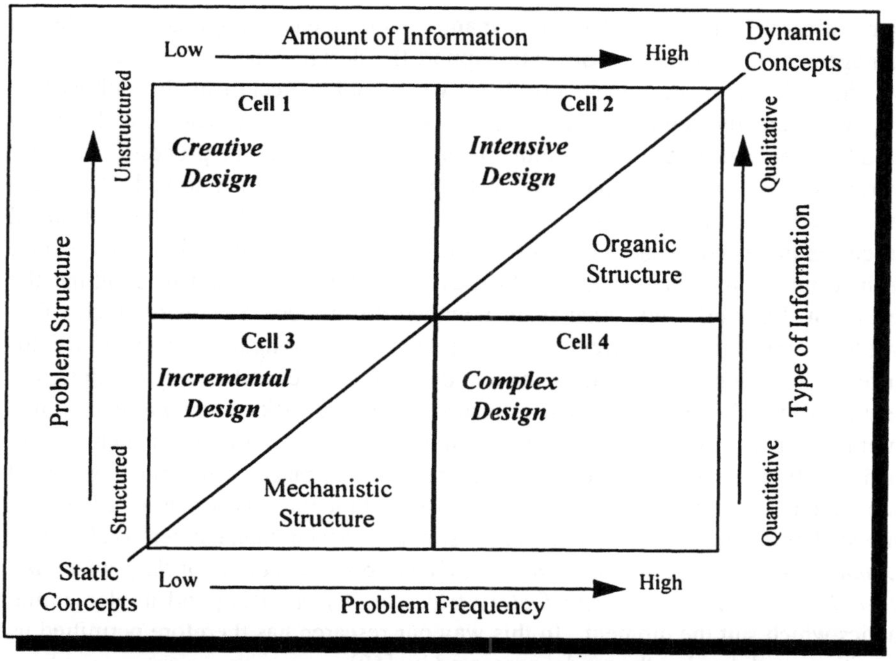

Figure 7. Contingency model of engineering design management (after [50])

There are two main criticisms of [50]. Firstly, their classification of designs involving either static or dynamic concepts is too simplistic. A viewed shared by [51] who rightly suggest that designs will contain both static and dynamic concepts, but to varying levels. The model is a classificatory scheme, designed as a typology to allow one to slot design problems into boxes, rather than attend closely to the design process and moment-to-moment shifts in task demands and appropriate management responses. Secondly, [50] describe the process of managing design as a process of managing information flows. However, as our work and others (see [4]) show, design management must also involve the management of people and their interactions.

If we open up the [50] model to create a three-dimensional space by adding a design team integration dimension (varying from low integration-high task orientation to high integration-low task orientation), a more complex response envelope becomes possible. Our task then becomes one of characterising the management structures associated with the various combinations of these settings.

Pugh (1996) defines Dynamic and Static as boundaries of a design continuum dependent upon the level of innovation involved in the design.

Our research indicates that a dynamic contingency approach to design management is necessary in order to support the design process. In section 3.2.2 we show that managers have problems which unlike technical difficulties that respond to creative solutions, are recurrent and sometimes intractable, and have to be addressed repeatedly throughout the product development process. In other words, they are problems that have to be managed rather than solved. Different tasks exert different kinds of demands, for example the different kinds of social processes that need to take place for a team to be effective. Additive tasks require each team member independently to make their best effort and to combine the results of this work with other team members to produce a team product. This would correspond to different team members being assigned responsibility for different subassemblies, working independently on these, and simply combing the results with only limited mutual adjustment. Other tasks exert more complex interactive demands. In particular, a disjunctive task requires team members to decide between alternative design solutions, and this places a premium on social interaction and mutual adjustment. If communication between engineers is impeded, this cannot be achieved. The task facing design management is *therefore to maintain a complex and dynamic equilibrium over the course of the project and steer the design team on a satisfactory trajectory* by choosing and implementing tactics which suit the moment. In this way our research has therefore permitted us to critique and develop the model presented by [50].

We propose that a management structure should be adopted which encourages the most appropriate combination of coordination mechanisms. Thus the team environment and coordinating mechanisms are aligned to a given situation as characterised by the contingency factors mentioned above. We envisage that this approach could be applied in two ways. Firstly, at the planning stage it could be used *strategically*, at a static point to plan the project team type etc. Secondly, whilst the project progresses it could provide dynamic advice upon *tactical* shifts and impromptu structures, based upon contingency factors changing. By providing a guide to suitable team and social structures, and the appropriate coordinating mechanisms for a given project situation we can support the management of coordination.

This approach relies upon dynamically monitoring a variety of contingency factors. Section 5 outlines the design of a socio-technical system based on this approach.

5. Supporting the Solution of Problems in the Management and Coordination of Design

5.1 Types of Facility Required

The case has been made for a support system which provides the following types of facility :

- Software tools to provide support for activities such as risk management and project control;

- Databases and access mechanisms to support the provision of organisational knowledge. These would be seen as primarily user-driven facilities, although software agents would also have access to their data and possibly some degree of limited control;

- Proactive software agents to drive and coordinate the communications between tools, databases and users. These would not be strictly autonomous in that aims and motives would be provided in the system design. However, aims and motives may be able to evolve to some extent. The role of agents would be to undertake the continuous and fine-grained monitoring of the contingency factors, discussed in section 4, to an extent which is beyond the practical scope of human project leaders. Software agents would also be responsible for many of the small-grain communications between other software facilities and users. By using software agents the system would address the difficulties in the dynamic monitoring, control and communication necessitated by the complex interactivity of project management problems which are described by the dynamic contingency management model;

- Working practices which reflect first those social mechanisms which have proved to be effective in facilitating the design process, and secondly the availability of software support in the areas identified.

The outlined system is not envisaged as a rigid, prescriptive straight jacket into which all design projects must fit. Rather, it is seen as a flexible collection of tools, working practices, roles and process models supported by interacting, goal-driven software agents, surrounding a database of organisational knowledge. Our intention is that the system would be used selectively and adapted to meet the needs of particular projects. Figure 8 shows how the various elements of the system are integrated.

5.2 Prototype Socio-technical System

The issues which are being addressed by the prototype are as follows:

a) *Process models*

Project planners will have access to a battery of process models one of which would be selected for a particular project, the appropriate model being selected upon criteria such as the level of innovation, marketing strategy etc. The model will specify the requirements of components of the project such as the level of documentation and roles required, which would in turn, be reflected in the type of documentation template used and personnel assigned to the team.

b) *Specification of customers' requirements.*

Specification of customers' requirements must remain flexible in order to meet the needs of the market. However, what can be greatly clarified is the intent of the specification, that is to say, a clear statement of where it lies on the continuum from vague wish list to firm specification of requirements. This would require recognition in working practices, documentation, planning and process models. Clear understanding of intent is a vital prerequisite for effective risk management.

c) *Management of risk*

Techniques and tools for the positive management of risk as described by [34] and [32], can reduce the potential for the addition of unanticipated innovation, together with its associated harmful effects on project time scales. Risk management would be based on an evolving set of parameters and would therefore be dependent upon the capture of, and access to, previous project risk management information.

d) *Organisational knowledge*

Organizational knowledge can be better captured and understood :

Transactive memory

A database identifying the sources of particular knowledge in a company's transactive memory [52], would enable designers more easily to find whom they should be talking to in order to solve a particular problem. Additionally, it would help in the formation of project teams and resource allocation, identify knowledge gaps in the company, and assist in determining training needs. In this way the system would support not just access to organisational knowledge but also the management of that knowledge. Part of the system would rely upon the definition and maintenance of appropriate model roles.

Externally stored knowledge

An object oriented database of template-based documentation as described by [36], specified by the selected process model for the project being undertaken, would support the provision of, and access to, information required for design reuse.

e) *Roles*

By defining and making explicit as job descriptions the roles that are *de facto* carried out, or currently neglected in design teams, motivation for undertaking "supporting" roles may be improved due to the recognition of their value. In addition the development of project teams should be made easier. This provision would also support the transactive memory system described above.

f) *Project planning*

More appropriate process models, clear specification of intent, active risk management, and definition of roles would make estimation, management and control of the process of design more effective even using existing software tools such as Microsoft Project. Facilities such as transactive memory mapping, and improvements in the capture and accessibility of past project planning information should also improve the accuracy of estimates.

g) *Working practice*

Work is still proceeding on the systematic analysis of our ethnographic survey data in order to identify appropriate working practices to complement the proposed system.

h) *Software agent.*

On their own, software tools and databases remain passive. What is required is an active, integrated system in which communication and coordination between facilities takes place automatically where possible, but under the control and with the full participation of the members of the design team. A number of agents are envisaged to achieve this dynamic nature, at present our focus has been the following :

Project agent: To gather and represent through a unified interface, the required system components and information specified by the process model selected for a project, thus providing a 'process model' based interface to satisfy the project team's needs. For example, once a process model has been selected for a given project, an agent will be created responsible for identifying and retrieving that project's needs from the system. The agent will also measure and communicate progress of the project against the process model. In addition the agent could liaise with other project agents e.g. to monitor resource overlaps.

Contingency agent: Through monitoring predefined attributes within the system this agent is triggered by potential problem situations. Based upon the contingency theory discussed in section 5 the agent will either alert appropriate personnel of the situation or advise them of a suitable course of action.

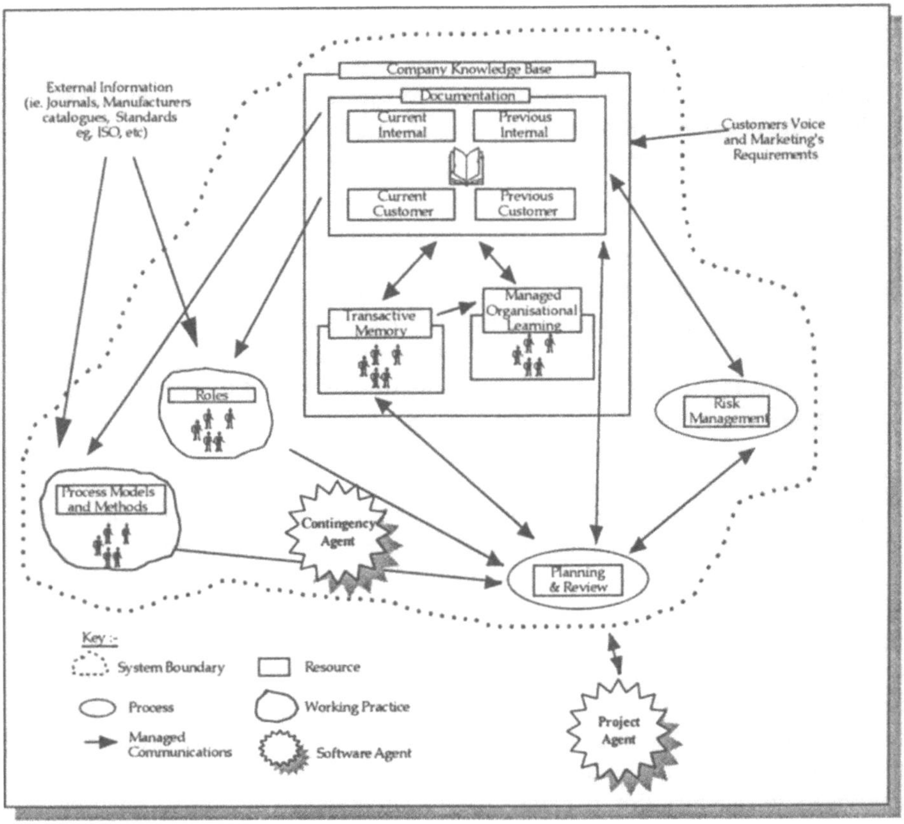

Figure 8. Socio-technical system infrastructure

6. PROSPECTIVE WORK

Following the completion of the final evaluation of the socio-technical model that has been discussed, key components of the system will be researched and developed into fully functioning components. This will require research and trials of the components in an industrial setting. Our industrial collaborator for this project is keen to progress the work and has agreed to further collaboration to meet these needs. It is envisaged that the key components will each form separate studies that will need to be conducted concurrently.

7. CONCLUSION

We have described a set of interdependent issues which our studies have shown to detract from the efficiency of the early, conceptual stages of the management and control of electronic engineering design. These problems are compounded by the complex dynamics of project team working which are only rarely amenable to static management and control tactics. Our analysis has indicated that the socio-technical system in conjunction with the contingency management framework that we have outlined, would support the solution of these issues.

REFERENCES

1. Pahl, G & Beitz, W (1984) Engineering Design. Springer-Verlag, Berlin.

2. Culverhouse, P F, Bennett, J P & Hughes, D R (1991 (a)) The EDT Electronic Product design process model. SERC/ACME report, 25/6/91.

3. Culverhouse, P F, Bennett, J P, & Hughes, D R (1991 (b)) A review of recent CAD/CAE tool developments.
SERC/ACME report, 15/4/91.

4. Cross, N & Cross A C (1995) Observations of teamwork and social processes in design. Design Studies, 16, 3.

5. Jenkins, D (1996) Supporting concurrent engineering in the age of teleworking and job sharing. Proceedings of Concurrent engineering and electronic design automation 1996, pp. 31-36. Society for Computer Simulation publication, Istanbul, Turkey.

6. Ullman, D G, Dietterich, T G, & Staufer, L A (1988) A model of the mechanical design process based on empirical data. AI for Engineering design, analysis, and manufacturing, 2(1), pp. 33-52.

7. Greenfield, O (1996) Time to market - Semiconductors. Proceedings of Concurrent engineering and electronic design automation 1996, pp. 31-36. Society for Computer Simulation publication, Istanbul, Turkey.

8. Checkland, P (1981) Systems thinking, systems practice. Wiley, chichester.

9. Coad, P & Yourdon, E (1991) Object oriented analysis (second edition). Prentice-Hall.

10. HUSAT (1988) Human factors guidelines for the design of computer based systems. HUSAT research centre, Loughborough University of Technology.

11. Morris, D, Evans, D, Green, P & Theaker C (1996) Object oriented computer systems engineering. Springer-Verlag, London.

12. Siddiqi, J & Chandra Shekaren, M (1996) Requirements engineering: the emerging wisdom. IEEE Software, March, pp. 15-19.

13. Jarke, M & Pohl, K (1994) Requirements engineering in 2001: (virtually) managing a changing reality. Software engineering, Nov, pp. 257-266.

14. Potts, C, Takahashi, K & Anton, A (1994) Inquiry based requirements analysis. IEEE Software, March, pp. 24-32.

15. Jackson, M (1995) Software requirements & specification. Addison-Wesley.

16. Goguen, J (1996) Formality & informality in requirements engineering. In proceedings of IEEE international conference on requirements engineering. IEEE CS press.

17. Davies, R M G & Saunders, R G (1988) Applying systems theory to project management problems. Project managment, vol 6, no 1, pp. 19-26.

18. Trist, E (1981) The evolution of socio-technical sytems. A series of occasional papers no. 2, from issues in the quality of working life. Ontario ministry of labour, Ontario quality of working life centre.

19. Schmidt, K & Bannon, L (1992) Taking CSCW seriously- Supporting articulation work. CSCW,1, pp 7-40.

20. Jirotka, M, Gilbert, N & Luff, P (1992) On the social organisation of organisation. CSCW, 1, pp. 95-118.

21. Sommerville, I, Rodden, T, Sawyer, P & Bentley, R (1992) Sociologists can be surprisingly useful in interactive systems design. In Monk, A, Diaper, D & Harrrison, M D (eds) People & Computers VII: proceedings of HCI'92, York 9/92. Cambridge Uni press.

22. Hughes, J A (1995) Ethnography, plans and software engineering. Proceedings of CSCW and the Software process, IEE Colloquium (2/1995).

23. Hughes, J, King, V, Rodden, T & Anderson, H (1994) The role of ethnography in interactive systems design. In proceedings of CSCW'94 conference on computer supported cooperative work. ACM, North Carolina.

24. Bansler, J P & Bodker, K (1993) A reappraisal of structured analysis: design in an organizational context. ACM transactions on information systems, vol 11, no 2, April, pp. 165-193.

25. Boehm, B W (1981) Software engineering economics. Prentice-Hall, London.

26. Davies, P (1995) The thought that counts. New Scientist, 6 may 1995, pp 27-31.

27. Drexler, K E (1992) Engines of creation. Oxford University press.

28. Merridth, J R, Raturi, A, Amoako-Gyampah, K & Kaplan, B (1989) Alternative research paradigms in operations. Journal of operations management, vol. 8, no. 4, pp 297-327.

29. Polson, P G, Lewis, C, Rieman, J & Wharton, C (1992) Cognitive walkthroughs: a method for theory based evaluation of user interfaces. International journal of man-machine studies, 36, pp.741-773.

30. Nielsen, J & Molich, R (1990) Heuristic evaluation of user interfaces. In proceedings of CHI'90 conference on Human factors in computer systems, pp.249-256. ACM, New York.

31. Boehm, B W (1984) Software Engineering Economics. IEE transactions on software engineering, vol 10, no 1, pp 5-21.

32. Gero, J S (1990) Design Prototypes: a knowledge representation schema for design. AI magazine; Winter, pp 27-36.

33. Henderson, R M & Clark, K B (1990) Architectural innovation: the reconfiguration of existing product technologies and the failure of established firms. Administrative Science Quarterly, 35, pp 9-30.

34. Culverhouse, P F (1993) Four design routes in electronics engineering product development. Journal of Design and Manufacturing, 3, pp 147-158.

35. Wegner, D (1986) Transactive Memory: a contemporary analysis of group mind. In Mullen, B & Goetals, G R (eds) Theories of Group Behaviour. Springer-Verlag, Berlin.

36. Culverhouse, P F (1995) Product Books: Archiving for information use and reuse, a replacement for CAD databases. Journal of Design and Manufacturing, 1, pp 1-16.

37. Suchman, L A (1987) Plans & situated actions: the problem of human-machine communication. Cambridge university press.

38. Parsons, T & Bales, R F (1964) Family: socialization & interaction processes. Routledge & Kegan Paul Ltd, London.

39. Belbin, R M (1993) Team roles at work. Butterworth-Heinemann.

40. Litterer, J A (1973) The analysis of organisations. John Wiley & Sons, New York.

41. Mintzberg, H (1983) Structure in Five's: designing effective organisations. Prentice-hall international, New York.

42. Burns, T & Stalker, G M (1961) The management of innovation. Tavistock publications.

43. Morgan, G (1986) Images of organization. Sage publications inc.

44. Galbraith, J. (1977) Organizational Design. Addison-Wesley.

45. Chandler, A D (1962) Strategy and structure. MIT press.

46. Fry, W F & Slocum, J W (1984) Technology, structure and work group effectiveness : a test of a contingency model. Academy of management journal, vol 27, no. 2, pp. 221-246.

47. Bays, M (1994) Organizing for information systems quality : a structural contingency theory investigation. PhD thesis, Monmouth college, New Jersey.

48. Van de Ven, A Delbecq, A & Koenig, R (1976) Determinants of modes of coordination within organizations. American sociological review, 41, pp. 332-338.

49. Argote, L (1982) Input uncertainty and organizational coordination in hospital emergency units. Administrative science quarterly, 27, pp. 420-434.

50. Slusher, E A, Ebert, R J & Ragsdell, K M (1989) Contingency management of engineering design. In proceedings of Institution of mechanical engineering international conference on engineering design, vol 1, pp. 65-75. Published by Institution of mechanical engineering, Bury St Edmonds.

51. Hosking, D & Morley, I E (1991) A social psychology of organizing. Harvester-Wheatsheaf.

52. Brown, P E S, Jagodzinski, A P & Reid, F(1995) Applying ideas of transactive memory to the development of IT support systems for design engineers. HCSD report 95/6. School of computing, University of Plymouth.

The Design Co-ordination Framework: key elements for effective product development

M M Andreasen
Dept. of Control and Engineering Design,
Technical University of Denmark, DK-2800 Lyngby, DK.

A H B Duffy and K J MacCallum
CAD Centre, University of Strathclyde, 75 Montrose Street,
Glasgow G1 1XJ, UK.

J Bowen
Dept. of Computer Science, UCC, College Road, Cork, Ireland

T Storm
Laboratory of Flexible Production Automation,
Delft University of Technology, Delft, NL

Abstract

This paper proposes a Design Co-ordination Framework (DCF) i.e. a concept for an ideal DC system with the abilities to support co-ordination of various complex aspects of product development. A set of frames, modelling key elements of co-ordination, which reflect the states of design, plans, organisation, allocations, tasks etc. during the design process, has been identified. Each frame is explained and the co-ordination, i.e. the management of the links between these frames, is presented, based upon characteristic DC situations in industry. It is concluded that while the DCF provides a basis for our research efforts into enhancing the product development process there is still considerable work and development required before it can adequately reflect and support Design Co-ordination.

1. Introduction

Competitive advantage can be achieved by a number of different means [1]. Within the current market environment, enterprises are being required to become increasingly responsive to changing and diverse customer needs, while being able to introduce and deliver their products more efficiently and at competitive prices. That is, manufacturing enterprises need to ensure:

- shorter time to market,
- improved quality and customer satisfaction, and
- competitive costs.

Over the past decade industrialists and academics have recognised the need and importance of developing approaches to enhance competitive advantage within manufacturing companies [2-14]. For example approaches such as: *Enterprise Modelling* which focuses upon the effective utilisation of the material use and manufacturing process within a company's business environment and the total integration of the enterprise [15, 16]; and *Concurrent Engineering* (CE) which has been primarily directed at ways of performing tasks in parallel and has resulted in a number of methods being developed as suggested by Vasilash [17], such as simultaneous engineering, life cycle engineering, process driven design, team approach, and design for manufacture.

In particular, the significance of the design process in determining the success or failure of a product in the market place and hence influencing a company's market strength is becoming increasingly more articulated [1, 18-20]. It is becoming more apparent that there is considerable scope for substantial improvements and efficiency gains within design. A survey carried out in the mid-80s suggested that overall `effectiveness' of development engineering is around 4% [21]. The basis for this statistic is that, on average, engineers reported spending around one third of their time doing "real" design, of this only one third was spent solving the "right" problems, and of this management have the right competencies for only one third of the time (i.e. $1/3 \times 1/3 \times 1/3 \approx 4\%$). That is, a considerable amount of time and effort is wasted by the lack of focus on the application and management of design effort. This leads to the conclusion that the potential for improvement in better productive use of engineering design resources is substantial - provided we have the mechanisms to realise it.

Unfortunately Enterprise Modelling approaches do not directly address the issues prevalent within Product Design Development and hence do not provide a means or foundation upon which to optimise the design process. These approaches have tended to concentrate upon global company functioning, strategy, overall business development and have tended to be directed at providing a high level director/managers' tool. Thus a major weakness of these approaches is that they do not directly address the issues involved in the actual design activity of complex and multi-disciplinary design projects.

An indication of the complexity of formalising the design process comes from the large number of descriptions which exist in the literature and which attempt to represent the essence of design [22]. What is interesting about these descriptions is their variety as well as the fact that they emphasise different aspects, according to the interpretation or findings of the author(s). They have also tended to model design as a chain of activities and have not taken into account the many complexities [23], life phase aspects [24] and business/enterprise issues [25] which are essential considerations for conducting effective design. Consequently, in recent years we have seen a trend which moves away from these "traditional" models towards "Concurrent Engineering".

CE research represents considerable global effort to shorten the design cycle time, improve the design quality, or reduce product costs. Within Western Europe the emphasis has been upon approaches which consider the product's *life cycle* issues [26-29] and the total integration of the market, design, and production functions of a business [30]. Thus, European work has tended to concentrate upon shortening the design cycle time by introducing and addressing life cycle issues earlier in the design process in order to alleviate problems and define a better quality of product by way of an integrated development process. In the United States of America (USA), more emphasis has been placed upon *co-operative working* as typified by the DICE (DARPA Initiative in Concurrent Engineering) program [8, 31] workshops [32], books such as that entitled 'Computer-Aided Co-operative Product Development' [33], and research publications [34-37]. In addition, supporting *information sharing* [38, 39] and *design management* [40-45] are currently topical issues in American CE research. While the research effort in Europe has tended to concentrate upon life cycle issues, the USA emphasis has primarily been upon enhancing integration and communications of human and computer resources. However the ethos in both continents has been one of attempting to follow the CE philosophy.

The design of complex products involves the co-ordinated organisation of multi-disciplinary groups, activities and information which continually evolve and change during the design process. Historically, different activities in the design process have been kept separate and interact through informal and formal communications, via paper medium or file transfers, resulting in considerable effort going into the resolution of inconsistencies, conflicts, and uncoordinated design activities. The design process is becoming increasingly and inextricably linked to computational design tools and data models which invariably remain disparate. Before multi-design tasks can be effectively organised and co-ordinated within such a computer medium, there is a need to integrate the data of the product model and the design tools in order to make effective use of the resources available. Such an environment can provide not only a basis upon which to develop multi-disciplinary design and dynamic project team organisations but also consistent, co-ordinated and efficient product development.

A major shortcoming of the Concurrent Engineering view is the failure to recognise that what is truly required is not for activities to be carried out in parallel but for resources to be effectively utilised in order to carry out tasks for the right reasons, at the right time, to meet the right requirements and give the right results. That is:

> *the key to achieving optimal design performance, and hence design productivity, is the effective co-ordination of the design process.*

The authors of this paper have been involved in a working group of researchers (CIMDEV) which was funded by the Commission of the European Communities

ESPRIT initiative from September 1992-1995, and subsequently as the Integration in Manufacturing and Beyond (IIMB) group from September 1996-1999. The group was initially formed in order to foster and develop collaborative links in basic research for DEVices (i.e. software tools) for Computer Integrated Manufacture (CIM). Since its inception, a sub-group of CIMDEV has played a leading role in the development of a key issue in design research - *Design Co-ordination* (DC) - which has now been adopted by the wider European group as a unifying concept for several areas of research, each examining different aspects of this central theme [46].

The argument for Design Co-ordination (DC) is that, to optimise design, activities should not necessarily be carried out "concurrently" but should be structured in such a fashion as to achieve optimum performance (such as total life quality and costs, cycle times, profitability, etc.). To achieve this, Design Co-ordination focuses upon issues directly relevant to the optimisation of the design process. Thus, the DC sub-group has formulated the following *mission* for the project:

- to achieve a quantum leap in the performance of the Product Development Process.

With the *principal goal*:

- to develop a computer based environment which supports the effective utilisation and integration of resources in order to optimise the design process.

This paper presents a framework for the effective co-ordination of the product design development process, the Design Co-ordination Framework (DCF). It is based upon the cumulative experience of the authors working in the field and discussions with a number of leading manufacturing companies. It is a hypothesis of the key elements involved in Design Co-ordination and as such forms the basis for discussion and a foundation to focus our research activities. Within the paper, before describing the DCF, a general outline of Design Co-ordination is given. Mechanisms for Design Co-ordination are then presented and the use of the framework in design discussed. The paper concludes the DCF it requires more thorough validation and will no doubt be subject to change and evolution.

2. Design Co-ordination

Design co-ordination may be seen as the activity of covering the need for co-ordinating the design activity in itself due to its own complexity, and the need for co-ordinating within complexities of activities, resources, goals and tasks of the company.

Design may be viewed as human and creative *problem solving* and *engineering* activities, market and commercial activities of *product development*, and planning and strategy related management activities of *product planning*. Co-ordination should bring these activities into a total organisational and commercial context [47].

The complexities which design co-ordination should cope with have been identified in an earlier publication [48]: the complexity of the artefact, the design activity, decision making, actors, aspects, knowledge and sources. Design Co-ordination brings these parameters together and allows for the management and control of the Co-ordination. One pattern of co-ordination is today well known and has high effects: concurrent product development. Concurrency is obtained by co-ordinating in accordance with creation of simultaneity, integration and providence of the activities of designing [48].

Establishing an interactive Design Co-ordination system means to manipulate important design factors like tasks, resources and design aspects related to the time dimension of designing. Design Co-ordination may therefore be seen as four main activities: decision making, controlling, modelling and planning/scheduling, with respect to the mentioned factors, see Table 1 [48].

Activities / **Factors**	Decision making	Controlling	Modelling	Planning/ Scheduling
Time				√
Aspects			√	
Resources		√		
Tasks	√			

Table 1. Activities of design co-ordination related to design factors [48].

The complexities of the design activity mentioned above are dimensions or parameters, which change state and structure during the design activity. The artefact to be designed grows in complexity, gradually new resources are brought in, new tasks are planned and a long row of different models support the synthesis, analysis and documentation activities. Decision making plays a central role, relating the organisation members with more or less influence on the artefact and the activities.

3. Key Elements - A Set Of Frames For Design Co-ordination

If we want to show or monitor the design activity and keep track of all the important relations between the activity and the company, what key elements do we need? In this section we try to identify this set of frames, where a frame is a model, showing the change of state or the change of relations.

Figure 1 shows a proposal for a set of frames for product development. The frames reflect the authors and a number of industrialists believe to be the most important aspects, but make no claim on completeness or correctness.

Each frame represents a particular aspect of design co-ordination, their state, evolution and relations. They represent models of different aspects of design. These models are not as yet matured in research or practise, because their role has been mind-setting or explanatory rather than as explicit tools. Thus, some of the models are new and find little support in the literature.

3.1 Model of Product Development - Frame 1

Product development is the activity which links need recognition to the introduction to the market place. There exists many theoretical, normative models in the literature (Eekels & Rozenberg [49], Pahl & Beitz [50], Hollins & Pugh [51], Andreasen & Hein [30] and many more) and many companies use templates or master procedures for their design activity (e.g. Lucas [52], Philips [53], Stork [54]). The models are given many names, which adds to confusion: Total Design [55], Product Introduction Process (PIP) [52], Integrated Product Development [30, 56], Product Life Cycle [57].

The content of such models differs widely, from simple problem solving models, via engineering design focused models, to models showing interrelated functional activities of a company, leading to establishing a new business.

The nature of the models differ from theoretical (principal) and generic models, procedural models (fitted to the company's normal standard activity, defining standard sub-activities and their sequences and relations), master plans (with detailed planning schemes, but not fitted to the actual project) and actual plans on different levels of scope and detail, i.e. showing the project strategy, the tactics or the operational activity pattern.

Such models or procedures may be related to a company handbook or manual, explaining each activity (input, output, definition, responsibility, methods, tools etc.). The handbook may be structured in accordance with the companies quality handbook, ISO 9000, or similar regulations.

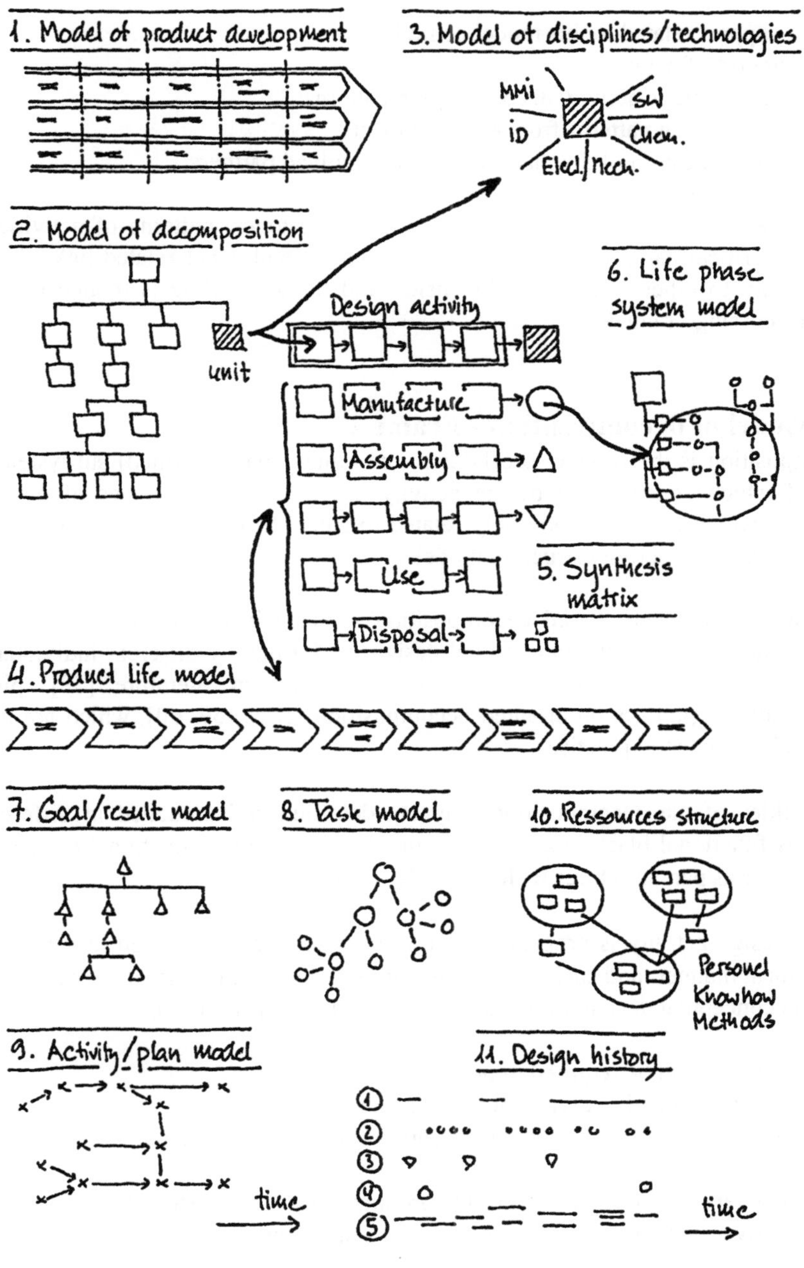

Figure 1: Set of frames for design co-ordination. Each frame symbolises a monitoring model. Co-ordination means establishing, managing, and controlling proper dynamic relations and inter-action between these frames.

The DC support related to this model could be:

- support of planning
- use of generic activity decomposition patterns
- reuse of plan patterns from earlier projects
- communication of the plan to different involved partners

The frame would show the activity elements, milestones, patterns of activities, relations, critical path, actors or agents involvement and event related links from one activity to others. The time dimension and the scheduling is treated below when discussing Frame 9.

3.2 Model of decomposition - Frame 2

Decomposition is the product breakdown into (functionally defined) subsystems, each of which constitutes a design activity. There exist other types of design activities, which cannot be precisely related to subsystems, so the decomposition model is not a full activity structure model.

Pragmatic, experience based decomposition is made in many companies, as a basis for activity planning and parallel design, performed by different individuals or teams. Here there are only immature theories in literature [58], plus mathematical, optimisation oriented decomposition styles [59, 60], which are not applicable for general practise.

The entities of the decomposition may be identified and modelled in different ways: as functional black-boxes, as structured elements with functional or spatial relations, or as part structures with assembly relations [61].

As the design activity is progressing, new subsystems are identified in a mainly hierarchical manner, and each subsystem is transformed from functional identification, via conceptual design, into a detailed part structure (Frame 5).

Such a model could be required to provide:

- a definition and overview of product subsystems
- an overview of alternative design routes and alternative solutions to subsystems
- an overview of reuse areas of the product (known subsystems)
- an overview of design progress of each subsystem.

The frame is a hierarchical structure of units, where each unit represents a subsystem by its identity, degree of classification, and where you have the possibility to see an illustration (e.g. CAD drawing) of each subsystem, when requested.

3.3 Model of disciplines and technologies - Frame 3

Product development changes technologies and know-how into new business possibilities. The R&D organisation has to gain new insights and ideas for use in the design activity. The R&D organisation have to be multidisciplinary in accordance with the needs of the product for creating mechanical, electronic or software solutions or to create a control circuit, a man/machine interface, optimisation of a fluid flow system etc.

It is necessary to divide the tasks of the design activity in accordance with the disciplines, but also to integrate those disciplines in such a way as to achieve overall optimal solutions. Seen from the manager of the design department it is necessary to create continuous and professional development of each discipline while the design projects need a limited professional contribution, e.g. for solving a software problem related to some logic of the man/machine interface.

At present we see the discipline model as a "discipline content and distribution" model of the artefact to be designed, but also the disciplines and expertise required to develop the design. It is recognised however that this model needs considerably further clarification and development

3.4 Product life model - Frame 4

Designing is closely linked to foreseeing product life phases, primarily the "use" phases, but also establishment, maintenance and liquidation. The perception of these phases or the product life scenario may be rather concrete and sure, due to familiarity of existing products, or unsure and unknown. But in any case the designer's task is to fit the product to an imagined life scenario.

Each life phase may be seen as a system; the product interacts with this system and the effectiveness of that interaction determines the performance and ease of performing/surviving the actual phase. Such life phase systems could be production, sales, transport, service or recycling systems [24]. The product specification normally defines the range and focus of life concerns and raises the need for a product life model for monitoring this range and focus and the need for relating product life facts to the design activity.

The functionality related to a product life model could be to:

- support for defining or identifying product life range and focus
- support for reuse of product life scenario and data from familiar, earlier products, established on the market place
- overview of design progress related to product life concern
- monitoring of product life aspects like cost, quality and environmental effects.

The monitoring of this frame could be supported by symbolic life phase/life cycle models, related data views or focusing on selected life aspects like total life costs [62, 63].

3.5 Synthesis Matrix model - Frame 5

The synthesis of the artefact is the core of the product development activity. Each subsystem or larger parts of the product may be treated in a separate design activity. The design activity may be unstructured or captured in steps with defined phase results like, for example, a concept. The literature and industrial practise offer many different models of the design activity, more or less fitted to product types and the type of industry, see Frame 1.

Adopting the concept of Concurrent Engineering we propose a parallel product and production engineering process [30, 53, 64]. The phases are synchronised and performed interactively and there are two results: a product and a production result. The content of the production engineering activity may range from a fully developed production system (e.g. a new automated assembly cell) to a quantitative fitting (selection of tools, jigs and production flow pattern) of the production system to the product.

Expanding this idea to other product life phases reveals a similar concept of parallel synthesis. The range of life phases should be in accordance with the range and focus of the product life model, see Frame 4. The output of each activity is a definition of the life phase conditions and/or systems. Thus, not only should issues involved in DFX be considered but also the X element but must also be designed and evolved along with the product. That is, we must Design The X (DTX) in order to ensure the overall business and product requirements are met.

The use of the synthesis matrix, i.e. the realisation of concurrent synthesis, could be the following:

- support for planning
- support for reuse of plan patterns
- support for reuse of concurrency patterns
- monitoring of parallel design progress
- communication of the plan and progress to relevant participants (e.g. partners or sub-contractors), showing updates and changes.

The frame representing the synthesis matrix could be a symbolic diagram or a principal activity relation network, with links to the results or to intermediate phase results and links between such results.

3.6 Life phase system model - Frame 6

As mentioned above, the concurrent system, identification or quantification of a life phase system has as a result a contribution to the specification of the life phase system and the life phase conditions. The responsible agents for the life phase (like the fabrication, assembly, transport or service department) need to receive these explicit results for control, adjustment or as a specification for the future life phase activity, like input for production preparation.

The need related to life phase system modelling is to model the different contributions from the synthesis activities and to transfer these contributions to the actual agent in their professional language. An example could be the transfer of chosen assembly methods, sequences, gripper surface information etc. to the assembly department, for contribution to the production preparation and the establishment of the assembly system.

For example, the assembly department may use this model to:

- obtain insight to assembly related design results
- control the development of the life phase system
- evaluate consequences on considerations such as cost and lead time
- allow reaction (feedback) to the design activity
- allow further detailing and progression in production planning.

The number of agents "receiving" this type of monitoring should be in accordance with the main focus of the design activity, mirrored in the product life model, see Frame 4. Some agents are not departmental, like installation and recycling, and in these cases the frame would serve the purposes of the design team.

The frame could reflect the assembly sequence contributions for each synthesised unit and the system layout contribution. Many life cycle areas have not developed characteristic models which again is an area requiring further consideration and development.

3.7 Product development goal/result model - Frame 7

The product development activity is normally controlled by goal specifications. The specifications define demands and optimisation criteria and specific required elements or features of the solution, with the basic idea to capture the properties of the ideal solution covering the perceived need.

Ideally seen this overall specification should be broken down into subsystem specifications as the design activity progresses, allowing evaluation and selection of best solutions for the subsystem. This breakdown should also ensure that subsystems in a proper way contribute to the overall optimisation of the total system.

State of the art of this area is not in accordance with this ideal situation. In some branches of industry specifications are not used. In more developed companies you find the breakdown of specifications only for sub-supplier deliveries or for choice of materials or units like motors, gears etc. The QFD method (Quality Function Deployment [65]) requires the breakdown of technical specifications from the total system via components to production processes and control operations, but no theoretical basis exists. The existence of a goal structure is foreseen by Ropohl [66] and some theoretical steps have been made [58].

As the design activity is progressing, the specifications are met and results may be formulated, i.e. a solution specification may be formulated. This specification reflects the decomposition model, and as the design progresses in detail, the results sum-up to overall performance parameters of the system [58].

The goal structure relates to a degree the decomposition structure. The specification elements (demands) cannot be related to the system elements beforehand, unless the solution is already well known. So the relation between the specification and the system elements is normally very complicated.

The aspects of DC that this model could provide support for are to:

- monitor the gradual elaboration of sub-specifications
- monitor the satisfaction of specifications and the obtained system performance
- control the results of the design activity at its milestones
- allow dynamic adjustments and additions.

The goal/result model is an overview of the specification elements and their relations to the decomposition structure. Each element within the goal structure could reflect a full specification document and also show the obtained resulting values.

3.8 Product development task model - Frame 8

The tasks of the team manager and the product development team may be formulated in a business specification which defines the purpose and goal of a project seen from a business point of view and act as a contract between the management and the team.

Also individuals and functional units of the basic organisation has formulated tasks to fulfil. Due to the nature of product development, where each new product changes the conditions for all functional units, these units also have specific tasks which are required to be performed in order to develop the product. Tasks of this nature could be cost reduction, service rationalisation, reduction of order treatment lead time, creation of data links to R&D, internal standardisation, ISO 9000 certification etc.

The defined tasks are more or less directly related to the product development activity and the solutions may be complicated: When do I, as service manager, interfere with the design activity for obtaining higher quality of the service departments activities?

Monitoring the task structure is complicated, but rewarding. There exist no theory or practise, but we believe that the transparency created by a Design Co-ordination system could allow the relations between the tasks and the product development activities to be explicitly expressed. In this way important links between product development and the basic organisation is established and may be managed.

3.9 Activity model - Frame 9

As outlined in Frame 1, many companies use a general master model of the product development activity as a basis for establishing a plan for an actual project. Where the model of the product development activity mentioned in Frame 1 serves the overview and monitoring of a set of activities (mainly showing strategy and tactics pius milestone actions), the activity model serves the control of teams and individual activities.

The DC support of an activity model could be the following:

- support the creation and maintenance of the project plan under dynamic changes
- support the reuse of plans from earlier projects
- communication of a common plan to all teams and agents involved in the project.

3.10 Resource model - Frame 10

An important task in design management is the allocation of resources. What are allocated are knowledge, skills and methods, carried by individuals, teams and equipment.

There are no models of the resources structure in current use in practise. The resources do not follow a hierarchy, they more likely reflect a network structure, where some of the nodes are agents outside the borders of the company.

The support from the resource model could be:

- identification and overview of resources and their network
- support for resource allocation and the utilisation of earlier resource allocations (experiences), related to the manager
- current and planned loading and allocations
- support for resource contributions seen from the agents point of view.

We consider that there is considerable effort and development required to adequately model the available resources, and support their effective utilisation, within the Design Co-ordination Framework (DCF).

3.11 Design history model - Frame 11

A design history model reflects central aspects of the product development activity, such as decisions and their rationale, related to product oriented solutions, design strategies, plans, allocations etc. The use of such a design history could be to:

- support documentation related to ISO 9000 etc.
- support reuse or any kind of insight obtained by previous projects
- allow learning from previous projects.

The design history model of the DCF is not in fact an active monitoring element, but more a passive register, with the ability to replay and reuse particular aspects of past design experience in order to optimise the co-ordination activity.

4. Mechanisms Of Design Co-ordination

In the previous section we proposed a set of frames, or monitoring models. We consider that the frames in themselves are important contributions for monitoring the design activity. In the following we will identify, how Design Co-ordination is related to these eleven frames. The frames themselves do not reflect co-ordination, but rather the elements involved in its support. Co-ordination relates to the effective control and management of the inter-relations between these frames, through appropriate DC mechanisms.

The need for co-ordination is related to the many complexities involved in design and the dynamic changes of design management parameters. Identifying the mechanisms of Design Co-ordination is partly a theoretical task, for setting up principal mechanisms, and a practise oriented task, finding those mechanisms which are important for an actual company in an actual situation.

A theoretical analysis could be an investigation of relations between frames created by dynamic changes in the design activity. We could ask for example: Where do we find the Design Co-ordination activity "decision making" in these dynamic patterns? In Design Co-ordination decisions are made, concerning artefacts, as part of synthesis and, about paths or "navigation" within the design activity. You find decision making on many levels, creating a continuum from strategic decisions down to detailed decisions of the designer. Therefore we find no strict demarcation of DC decision making.

Clarifying the decision making activities may be done in a matrix of the eleven frames, where it is noted, how a change in one frame could lead to decisions related to the same or other frames, see Figure 2. The result is a complex pattern, even if we only note higher level management oriented decisions, and the analysis does not tell us anything about the importance of the decisions and the need for co-ordination support

Effect on: / Cause here:		Product Development	Decomposition	Disciplines & Technologies	Product Life	Synthesis Matrix	Life phase systems	Goal/Results	Tasks	Activities/Plan	Resources	Design history
		1	2	3	4	5	6	7	8	9	10	11
Product Development	1	X	X	X	X	X	X	X	X	X	X	
Decomposition	2		X	X	X	X	X	X	X	X	X	
Disciplines & Technologies	3		X			X		X	X	X	X	
Product Life	4			X	X	X	X	X	X	X	X	
Synthesis Matrix	5		X		X	X	X	X		X	X	
Life phase systems	6						X	X		X	X	
Goal/Results	7			X	X	X				X	X	
Tasks	8	X		X	X			X	X	X	X	
Activities/Plan	9							X	X	X	X	
Resources	10	X		X				X	X	X		
Design History	11											

Figure 2: Design co-ordination occurs when changes in one frame (vertical list) propagate decisions about change in another frame (horizontal list). The crosses mark the principal decision related links.

Another approach, also theoretical, would be to set up a semantic net of dynamic changes. An example is shown in Figure 3, where two frames are related, namely the "Model of decomposition" and the "Synthesis matrix" with focus on a unit being designed. This analysis is very general and does not lead to a detailed definition for development in say a computer based system.

In Figure 4 is shown a similar analysis linking the "Model of decomposition", "Goal structure" and "Activity model" based on the identification of a new unit in the decomposition structure.

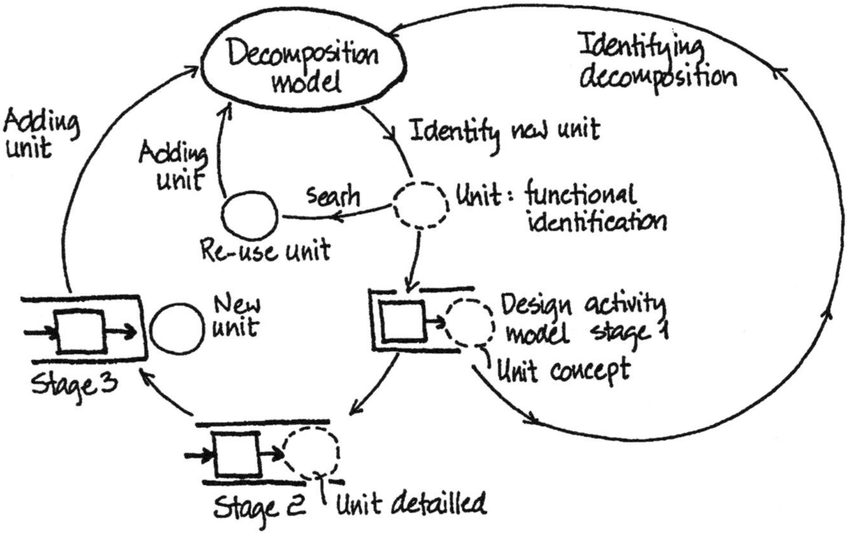

Figure 3 Example of analysis of dynamic changes in the frame network. Here two frames are related, namely the model of decomposition and the synthesis matrix model, with focus on the unit design.

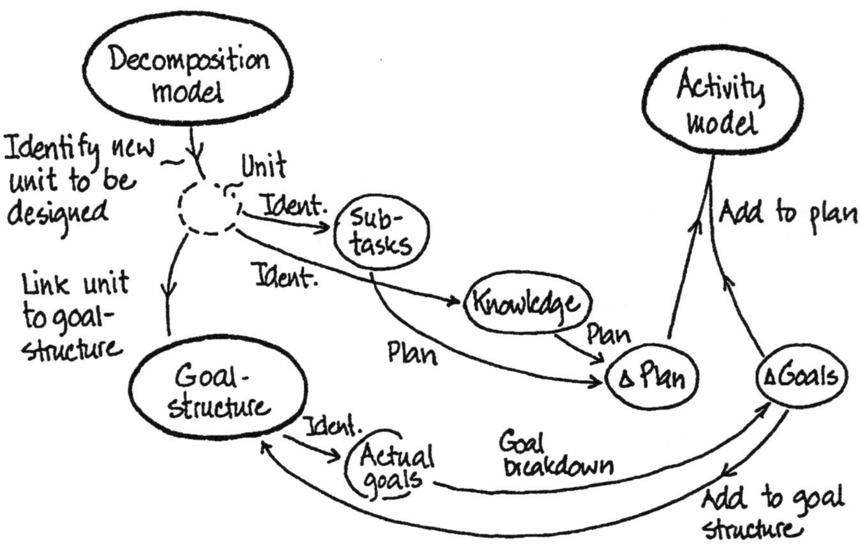

Figure 4 Example of analysis of dynamic changes in the frame network. Here the frames 'decomposition model', 'goal structure' and 'activity model' are related, due to an identification of a new unit in the decomposition structure.

A pragmatic/empirical approach would be to select a company, point out an actual project and to let the participants identify their co-ordination problems and the dynamic change-patterns in a prototype set-up of a computer based Design Co-ordination Support System (DCSS). This prototype could, as the first simple solution, be a white-board model of the frames and yellow "post-it" slips carrying the dynamic changes.

5. Discussion

Appropriate and fully developed mechanisms and frames for Design Co-ordination do not exist, either paper-based, in manual design today, or as computer-based design support systems. Thus, a critical condition for Design Co-ordination feasibility is that we are able to create frames which in themselves are of advantage to the execution and management of design.

The set of frames shown in Figure 1 reflect the design activity by the following frames: "Model of decomposition" and "Systems matrix model", not by automatic techniques, but as notes or indications from the design agents: Where are we? What are we working on? What results are obtained?

The set of frames is a neutral, generic, structure and gives only benefits, if it is utilised. It is well known from literature that a structured procedure and good use of earlier solutions both create high efficiency and quality results. Properly managed, the DCF could influence design in a very positive way, but will also change the nature of design, by prescriptive patterns, by external control of intermediate results, by monitoring progress against goals and tasks etc.

Some general principles concerning designing in a computer based DC Support System (DCSS) has been identified as:

- Any type of design support shall be accessible at any time (functional, conceptual or concrete). The DCSS should not dictate one design methodology but be flexible for individual types.
- It shall be possible to design the design approach. There are big differences in designers individual approaches: experimental, reuse, trial-and-error, mainly concrete or conceptual focused, with many/few alternatives, with individual graphic stile etc.
- The DCSS shall be integrated with existing and new information structures of the company.
- The DCSS needs to model the product structure (to be based on a product model) in order to aid the design process and Design Co-ordination.

7. Acknowledgements

The authors express their thanks to Dr S M Duffy for her work and contribution in Design Complexities and Team Engineering and wish to express their gratitude to the individuals and companies, who have joined the discussions and in this way have inspired and focused the work:

- M.L. Boerstra, Stork Demtec, NL
- P. Carr, Digital Scotland Ltd., UK
- Lars Hein, Institute of Product Development, DTU, Denmark
- D. Rimmer, Pilkington Optronics Ltd., UK
- Brian Sinclair, Mentec International, Ireland
- A. Travers, Scottish Design, UK
- Michael Vaag, Danfoss Ltd. Denmark
- Gert Yde, Bang & Olufsen A/S, Denmark

8. References

1. *Design for Competitive Advantage - Making the most of design.* in *International Conference on Design for Competitive Advantage.* 1994. Coventry UK: Institute of Mechanical Engineers.

2. Gregory, S.A. *Business, engineering design and management,.* in *Proceedings of the International Conference on Engineering Design Hamburgh 26-28 August 1985.* 1985. Heurista Zurich.

3. Chalmet, L.J. *Increasing your competitive advantage through CIM,.* in *Proceedings of the Third CIM Europe Conference Knutsford UK 19-21 May 1987.* 1987. IFS(Publications) Ltd UK /Springer-Verlag Berlin.

4. Deasley, P.J. and S.J. Williams. *Competive products through innovative control,.* in *Proceedings of the International Conference on Engineering Design Harrogate UK 22-25 August 1989.* 1989. Mechanical Engineering Publications Ltd.

5. Stenberg, H. *Design flexibility - a competive factor?,.* in *Proceedings of the International Conference on Engineering Design Harrogate UK August 22-25 1989.* 1989. Mechanical Engineering Publications Ltd.

6. Pennel, J.P., *et al. Concurrent Engineering: an overview for Autotestcon,.* in *IEEE Internation Automatic Testing Conference The Systems Readiness Technology Conference Philadelphia PA USA 25-28 September 1989.* 1989. \it IEEE Piscataway NJ USA.

7. Foreman, J.W. *Gaining competitive advantage by using simultaneous engineering to integrate your engineering, design and manufacturing resources,.* in *Conference Proceedings of Autofact Detroit Michigan 30 October-2 November 1989.* 1989.

8. Sprague, R.A., K.J. Singh, and R.T. Wood, *Concurrent engineering in product development,.* IEEE Design and Test of Computers, 1991. **8**(1): p. 6-13.

9. Jackson, S. and M. Romeri, *World class product development,*. Manufacturing Breakthrough: Managing the Design and Development Process, 1992. **1**(4): p. 233-238.
10. IT R&D Programme Computer-integrated manufacturing and engineering: Summaries of Esprit projects I and II - December 1993, ed. S. Rogers and M. Vereczkei. 1994, European Commision Directorate-General XIII Telecommunications Information Market and Exploitation of Research: Luxembourg.
11. Gilmore, D., *Keeping PACE with the market,*. World Class Design to Manufacture, 1994. **1**(1): p. 12-16.
12. *IFIP WG5.7 Working Conference on Managing Concurrent Manufacturing to Improve Industrial Performance, Seattle, Washington, USA, 11-15 September 1995,* .
13. *Manufacturing Breakthrough: Managing Product Development,* , Information for Success (IFS) Ltd Publishers Bedford UK.
14. *World Class Design to Manufacture,* , MBC University Press Ltd Bradford UK.
15. Roboam, M., *GRAI-IDEF-Merise (GIM): Integrated methodology to analyse and design manufacturing systems,*. Journal of Computer Integrated Manufacturing Systems, 1989. **2**(2): p. 82-89.
16. Kosanke, K., *The European approach for an Open System Architecture for CIM (CIM-OSA) - ESPRIT project 528 AMICE,*. Computing & Control Engineering Journal, 1991: p. 103-108.
17. Vasilash, G.S., *Simultaneous engineering: management's new competitiveness tool,*. Production, 1987. **99**(7): p. 36-41.
18. Coplin, J.F. *Engineering design - a powerful influence on the business success of manufacturing industry,*. in *Proceedings of the International Conference on Engineering Design Harrogate UK 22-25 August 1989*. 1989. Mechanical Engineering Publications Ltd.
19. *Improving Engineering Design: Designing for competitive advantage,*. 1991, Washington DC USA: National Research Council, National Academy Press.
20. Deisenroth, M.P., J.P. Terpenny, and O.K. Eyada. *Design for competitiveness: case studies developed in concurrent engineering and design for manufacturability,*. in *The Winter Annual Meeting of the American Society of Mechanical Engineers Anaheim CA USA 8-13 November 1992*. 1992. ASME New York NY USA.
21. Jaikumar, R., *Project Portfolios,* , Fourth International Forum on Technology Management Berlin Germany 18-20 October 1993.
22. Finger, S. and J.R. Dixon, *A review of research in mechanical engineering design. part 1: descriptive, prescriptive and computer based models of design processes,*. Research in Engineering Design, 1989. **1**: p. 51-71.
23. Duffy, S.M. *The Design Complexity Map and the Design Coordination Framework,*. in *Proceedings of the tenth Integrated Production Systems Research Seminar*. 1995. Fuglsoe, Denmark.

24. Olesen, J., *Concurrent development in manufacturing based on dispositional mechanisms,*, in *Institute for Engineering Design.* 1992, Technical University of Denmark: 2800 Lyngby.

25. Polito, J., A. Jones, and H. Grant, *Enterprise integration: A tool's perspective,*, in *Information and collaboration models of integration*, S.Y. Nof, Editor. 1994, Kluwer Academic Publishers: Dordrecht The Netherlands. p. 149-167.

26. Andreasen, M.M., *et al.*, *Design for Assembly,*. 1987: IPS Publications/Springer-Verlag London UK.

27. Tichem, M. *Design for manufacturing and assembly: a closed loop approach,*. in *Proceedings of the Ninth International Conference on Engineering Design The Hague 17-19 August 1993.* 1993. Heurista Switzerland.

28. Institute for Product, D., *Design for Manufacture: A guide for improving the manufacturability of industrial products,*. 1994.

29. Miles, B.L. and K.G. Swift, *Working together,*. Manufacturing Breakthrough, 1992. **1**(2).

30. Andreasen, M.M. and L. Hein, *Integrated Product Development,*. 1987: IFS (Publications) Ltd and Springer-Verlag London UK.

31. Ashley, S., *DARPA initiative in concurrent engineering,*. Mechanical Engineering, 1992: p. 54-57.

32. Sriram, D., R. Logcher, and S. Fukuda, eds. *Proceedings of the MIT-JSME Workshop on Cooperative Product Development,*. . 1989: Dept. of Civil Engineering, Massachusetts Institute of Technology, Cambridge, MA, USA.

33. *Computer-Aided Cooperative Product Development,*, , D. Sriram and R. Logcher, Editors. 1991, Springer-Verlag: New York USA.

34. Klein, M. and S.C.Y. Lu. *Insights into cooperative group design: experience with the LAN design system,*. in *Proceedings of the Sixth International Conference on Applications of Artificial Intelligence in Engineering.* 1991. University of Oxford UK.

35. Sriram, D., *et al.*, *DICE: An object-oriented programming environment for cooperative engineering design,*, in *Artificial Intelligence in Engineering Design: Knowledge acquisition commerical systems and integrated environments*, C. Tong and D. Sriram, Editors. 1992, Academic Press Inc.: San Diego CA USA. p. 303-366.

36. Visser, W. *Collective design: a cognitive analysis of cooperation in practice,*. in *Proceedings of the Ninth International Conference on Engineering Design The Hague 17-19 August 1993.* 1993. Heurista Switzerland.

37. Anupam, V. and C.L. Bajaj, *Shastra: multimedia collaborative design environment,*. IEEE Multimedia, 1994: p. 39-49.

38. McGuire, J.G., *et al.*, *SHADE: Technology for knowledge-based collaborative engineering,*. Journal of Concurrent Engineering: Applications and Research (CERC), 1993. **1**(3): p. 137-146.

39. Toye, G., *et al.*, *SHARE: A methodology and environment for collaborative product development,*, in *Proceedings of the 2nd Workshop on Enabling Technologies: Infrastructure for Collaborative Enterprises.* 1993, IEEE Computer Press. p. 22-47.

40. Kusiak, A. and K. Park, *Concurrent Engineering: decomposition and scheduling of design activities,*. International Journal of Production Research, 1990. **28**(10): p. 1883-1900.
41. Eppinger, S.D., *Model-based approaches to managing concurrent engineering,*. Journal of Engineering Design, 1991. **2**(4): p. 283-290.
42. Kusiak, A. and J. Wang, *Decomposition of the design process,*. Journal of Mechanical Design, 1993. **115**: p. 687-695.
43. Kusiak, A. and U. Belhe, *Scheduling design activities,*, in *Information and collaboration models of integration*, S.Y. Nof, Editor. 1994, Kluwer Academic Publishers: Dordrecht The Netherlands. p. 43-60.
44. Pourbabai, B. and M. Pecht, *Management of design activities in a concurrent engineering environment,*. International Journal of Production Research, 1994. **32**(4): p. 821-832.
45. Eppinger, S.D., *et al.*, *A model-based method for organising tasks in product development,*. Research in Engineering Design, 1994. **6**(1): p. 1-13.
46. *CIMDEV Working Group 7401,*, in *IT R&D Programme Basic Research: Summaries of Esprit projects working groups and networks of excellence - December 1993*, S. Rogers and M. Vereczkei, Editors. 1994, European Commision Directorate-General XIII Telecommunications Information Market and Exploitation of Research: Luxembourg. p. 404-407.
47. MacCallum, K.J., *Requirements for intelligent support of concurrent engineering,*, in *Artificial Intelligent in Design (AID'92) Workshop.* 1992: Carnegie-Mellon University, Pittsburgh, USA.
48. Duffy, A.H.B., *et al.*, *Design coordination for concurrent engineering,*. International Journal of Engineering Design, 1993. **4**(4): p. 251-265.
49. Roozenburg, N.F.M. and J. Eekels, *Product Design: Fundamentals and Methods.* Product Development - Planning, Designing, Engineering, ed. N. Cross and N. Roozenburg. 1995: John Wiley & Sons Ltd., Chichester Po19 1UD, United Kingdom.
50. Pahl, G. and W. Beitz, *Engineering Design: A systematic approach,*. English, 2nd ed, ed. K. Wallace. 1996, London: Springer.
51. Hollins, W.J. and S. Pugh, *Successful Product Design - What to do and When.* 1990, London: Butterworth.
52. Edwards, S. and P. Lewis, *The Product Introdcution Process.* 1991, Birmingham: Lucas Engineering & Systems Ltd.
53. Reinders, H., *From abstract idea until production,* . 1989, CTB Report 23/89EN, Centre for Manufacturing Technology, Philips Bedrijven BV: Eindhoven.
54. Minne, H.P. and M.L. Boerstra, *Computer Aided Design for Manufacture - a tool to be used at a designers workbench*, in *Proc. CIRP Conference.* 1990: Entschede.
55. Pugh, S., *Total Design: Integrated methods for successful product engineering,*. 1990, Wokingham England UK: Addison-Wesley Publishing Company.

56. Hananel, D., H. Hjort, and D. Lucas, *Customer Focused Product Planning & Implementation*, . 1991, Innovata Inc.: Cincinnati, Ohio, USA.
57. Sheldon, D.F., *Designing for whole life costs at the concept stage.* Journal of Engineering Design, 1990. **1**(2).
58. Svendsen, K.H., *Optimisation of composite mechanical systems*, in *Institute for Engineering Design.* 1994, Technical University of Denmark: 2800 Lyngby.
59. Azarm, S. and W.C. Li, *A two-lvel decomposition method for design optimisation.* Engineering Optimisation, 1988. **13**.
60. Sobieszczanski-Sobieski, J., *Interdisciplinary and MultiLevel Optimum Design.* Computer aided optimal design - structural and mechanical systems. 1987: Springer-Verlag.
61. Duffy, A.H.B., A. Persidis, and K.J. MacCallum, *NODES: a Numerical and Object based modelling system for conceptual engineering DESign.* Knowledge-Based Systems, 1996. **9**: p. 183-206.
62. Corbett, J., ed. *Design for Manufacture.*. 1991, Addison-Wesley: New York.
63. Perks, R., D. Sheldon, and S. Bush. *Whole life cost.* in *WDK Workshop - Design for X.* 1993. Institute for Engineering Design, Technical University of Denmark, 2800 Lyngby, DK.
64. Syan, C.S. and U. Menon, eds. *Concurrent Engineering - Concepts, Implementation and Practice.*. 1994, Chapman & Hall Ltd.: London.
65. Clausing, D. and S. Pugh. *Enhanced Quality Function Deployment.* in *Int. Conf. on Design and Productivity.* 1991. Honolulu, Hawai, USA.
66. Ropohl, G., *Flexible Fertingungs Systeme*, . 1971, Otto Krausskopf-Verlag GmbH: Mainz.

Part III
Concurrent Engineering

Concurrent Engineering: A Successful Example for Engineering Design Research

T Tomiyama
Department of Precision Machinery Engineering, The University of Tokyo
Hongo 7-3-1, Bunkyo-ku, Tokyo 113, Japan

Abstract

This paper reviews concurrent engineering from the viewpoint of engineering design research. Concurrent engineering aims at eliminating unnecessary changes and redesigns from a product development process, and at achieving better product quality. To do so, it is critical to facilitate teamwork and collaboration by improving communication among various participants in the development process and by allowing them to share information and knowledge. In contrast, the engineering design research community has worked on general and abstract theories that are too difficult to apply. To improve this situation, three recommendations for the design community are made. One is to focus on general but concrete problems. Second, research should be directed toward clear goals such as arriving at innovative designs more efficiently. Third, engineering design research should tackle design knowledge issues including ontology, models and model operations.

1. Introduction

Compared with some five years ago, the engineering design research community in general seems to suffer from the lack of funding and stagnation of innovative research results. Have a look at, for example, recent proceedings of the ASME's Design Theory and Methodology [1] which certainly lost momentum that could be observed some years ago. In 1995 at its annual conference, it even hosted a panel session entitled "Whither DTM" that inevitably evoked controversial discussions. Generally speaking, in many countries academic research in this field is now less funded than some years ago. In contrast, the manufacturing engineering research community that deals with design related topics slightly differently seems to suffer less.

Why does this situation happen to the design community that is considered the central activity in manufacturing? To answer this question, we look at concurrent engineering that the design community studies with sufficient interests. This paper will focus on concurrent engineering, because it is not only successful industrial practices but also concurrent engineering is a field largely related to design.

Historically speaking, " concurrent engineering can be seen as an American attempt to understand, implement, and surpass the best Japanese product-development practices" in the late 1980s [2]. It is a methodology to conduct a series of product development activities largely "concurrently," so that lead time for product development can be shortened and a variety of product aspects can be taken into consideration at an early design stage [3-6]. In the late 1980s and the beginning of 1990s, applications of concurrent engineering methodologies to product development were seriously pursued. These resulted in a variety of successful cases in aerospace, automobile, computer, semi-conductor, telecommunication, and home-appliance industries, to name a few.

The rest of the paper is organized as follows. The next section examines concurrent engineering first as industrial practices and second as academic research topics. It will analyze advantages of concurrent engineering but with some cautions to understand the intrinsic value of concurrent engineering. We then identify distinctive research directions within engineering design research community; i.e., "general vs. specific" and "abstract vs. concrete." These distinctions are vital to understand why concurrent engineering research was successful, while engineering design research was not. We contrast engineering design research with concurrent engineering research to draw three recommendations for future research in engineering design. One is to focus on general but concrete problems. Second, research should be directed toward clear goals such as arriving at innovative designs more efficiently. Third, engineering design research should tackle design knowledge issues including ontology, models and model operations.

2. Concurrent Engineering

2.1. Concurrent Engineering in Industry

Formerly known also as "simultaneous engineering," "collaborative design," or "cooperative product development," concurrent engineering is perhaps the most discussed and exercised industrial practice nowadays. Concurrent engineering was first explicitly practiced by the US industries who were motivated by the awareness about their competitiveness typically warned in [7]. A famous example of concurrent engineering is the development of the Boeing 777 commercial aircraft. The aircraft was designed and built by geographically distributed companies that worked entirely on a common product database of CATIA without building physical mock-ups but with digital product definitions. They further allowed airlines (i.e., customers) to participate and reflect their ideas in the development process, whose voices had been usually considered only after delivery.

Traditionally, a product development process was conducted sequentially. Usually, there is no good communication among these participants in the process and the whole process is so much haunted by Murphy. Consequently, this traditional pipeline process is inevitably characterized by numerous design changes and redesigning, wrong specifications, wrong design embodiment, unmanufacturable blueprints, intolerably late delivery of components, sudden discovery of unpurchasable components, loss of information in the long pipeline, and, thus, numerous trials and errors. The famous example of designing a swing tells us that what the customer wants is wrongly interpreted by the sales department, which produces specifications wrongly communicated to the design department, that impracticably designed the swing, which is incorrectly fixed by the manufacturing department [8].

Different from the traditional sequential product development process, concurrent engineering advocates subprocesses of design and development concurrently performed with or without help of advanced computing technologies. By doing so, lead time can be substantially shorter. This is possible not because individual development subprocesses become shorter, but because unnecessary changes were eliminated and infeasible specifications were detected so that re-designing could be avoided. In order to perform concurrent engineering in a real industrial setting, we can identify critical issues including:

(1) Building a cross-functional team exclusively focused on a target product, which boils down to also managerial and organizational issues.
(2) Facilitating mutual communication among participants of the cross-functional team, which implies that concurrent engineering aims at "automation of friendship."
(3) Bringing up considerations of later stages, such as manufacturing, purchasing, operation, maintenance, and recycling, to the discussion table.
(4) Developing a computational infrastructure to facilitate these three issues.

2.2. Advantages of Concurrent Engineering

Concurrent engineering is said to bring up with such advantages over traditional sequential processes for product development as better design productivity, more customer satisfaction, shorter lead time, and reduced development costs. However, these advantages seem misleading or even wrongly addressed, if we examine them in greater detail. In the following, we examine these "myths" of concurrent engineering. (Unfortunately, there is few published papers describing real data comparing product development processes with and without concurrent engineering. The followings were obtained from my personal communications with engineers and designers from the Japanese industry.)

(1) *Concurrent engineering improves design productivity.*
This depends on the definition of design productivity. If productivity is measured by "design output/lead time," concurrent engineering drastically improves design productivity. However, if productivity is measured by "design output/man-months," the improvement is not that significant. For instance, very typically, concurrent engineering methodologies necessitate to organize more meetings for design review involving a large number of participants including suppliers and even customers from an early design stage. Figure 1 compares a traditional

sequential product development process and one based on concurrent engineering. In an extreme case, while the total lead time becomes shorter due to elimination of unnecessary design changes and redesigns, concurrent engineering does not make individual subprocesses shorter. It must be restated that "*concurrent engineering improves design efficiency.*"

(2) Concurrent engineering improves customer satisfaction.
Certainly, fewer problems are found in later stage of the product development process, because considerations on more life cycle aspects are taken into the early design stage. In this sense, concurrent engineering can improve product quality which is one of the factors to achieve customer satisfaction. However, this does not mean more customer satisfaction. Unless explicitly taken into consideration, just concurrently performing product development activities does not automatically guarantee customer satisfaction. More importantly, taking into too many considerations may well end up with a "mediocre" product with "good" quality.

(3) Concurrent engineering makes lead time shorter.
Concurrent engineering can eliminate unnecessary changes and redesigns by which lead time can be shorter. However, it does not necessarily make individual development subprocesses shorter as discussed above.

(4) *Concurrent engineering cuts down development costs.*
This can be achieved only when unnecessary changes and redesigns are eliminated. Because concurrent engineering facilitates taking more aspects into consideration, it tends to mean more costs. However, note that good practices of concurrent engineering well cut down "production costs." This is typically observed in research in "Design for Manufacturability" and "Design for Disassemblability."

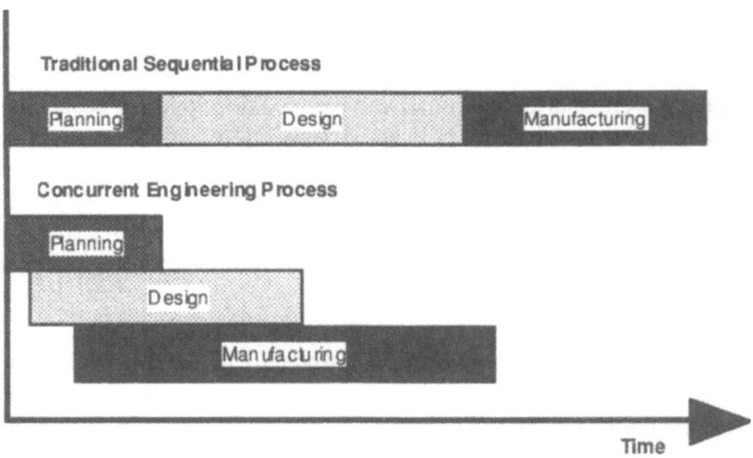

Fig. 1. *Traditional and concurrent engineering processes for product development. With concurrent engineering methodology, subprocesses of product development are largely concurrently performed, thereby arriving at shorter lead time than the traditional pipeline type process.*

Problems that concurrent engineering addresses arise due to excessive division of labor and due to too complicated products. In other words, when there was little division of labor, these concurrent engineering problems did not exist. It should be noted that, however, simply doing things concurrently does not improve anything. Concurrent engineering practices are successful, only by eliminating unnecessary changes and redesigns and by facilitating cross-functional teamwork based on mutual communication supported with integrated product life cycle knowledge. For example, just providing an integrated computational framework does not guarantee the success of concurrent engineering.

In addition, concurrent engineering cannot be performed without introducing sound technical philosophy that increases concurrence of product development subprocesses. Ward *et al.* [9], for instance, reports that Toyota employs a sophisticated method to incorporate "vaguely represented information" in the product development process.

These are critical observations that we should consider in defining research agenda.

2.3. Academic Concurrent Engineering Research

Research communities from a variety of disciplines, including engineering design, manufacturing engineering, managerial science, and computer science, started responding to the call for concurrent engineering, especially in the late 1980s. For researchers, the research issues of concurrent engineering are twofold. One is to incorporate a variety of life cycle issues (for instance, manufacturing, purchases, assembly, operation, maintenance, recycling, and disassembly) at an early stage of design [10]. The other is to develop computational tools and frameworks for coordinated product development activities [3-6]. Finger *et al.* [11] summarize the discrepancies between the industrial concurrent engineering practices and the academic research issues as follows :

For the engineering research community, concurrent engineering means, for the most part, the use of computational techniques to build cooperating sets of tools from different areas of design and manufacturing using specialized representations and coordination mechanisms. We call these technical aspects, which encompass engineering and computational issues. For industry, concurrent engineering has been interpreted as the creation of cross-functional teams that include people responsible for all aspects of the product life cycle. We call these organizational aspects, which encompass managerial, communication, and coordination aspects. While both approaches to concurrent engineering have the goal of information integration, the research has remained separate in the design literature.

We can further elaborate this observation. While the manufacturing research community focused primarily on Design for Manufacture from very early days

[12], the design community looked at the following four major research topics according to various publications [8].

(1) *Communication aspects* address the use of advanced computer-supported communication tools and techniques for knowledge and information sharing, such as CSCW (Computer Supported Cooperative Work) and multi-media technology based advanced computer networks.
(2) *Organizational aspects* include managerial issues such as building and coordinating a teamwork environment and enterprise integration, leading to cooperative work, better communication, and more responsive decision making processes.
(3) *Technological aspects* consider better communication and management including integration of various kinds of knowledge. Typical topics range from information modeling, such as product-process-enterprise modeling and simulation, data version control and management, and capturing design rationale, to computational frameworks and methods, such as integrated knowledge bases and databases, conflict resolution techniques, constraint management, planning and scheduling, and cooperative problem solving.
(4) *Life cycle aspects* that are more recent development including Design for X (DfX). In early days, X stood for almost exclusively manufacturing and assemblability, but more recently expanded to cover serviceability, maintainability, recycleability, disassemblability, reusability, and so on. Bringing up various kinds of knowledge about product life cycle into an early design stage is critical in product development processes.

3. Engineering Design Research and Concurrent Engineering

3.1 Research Dimensions

In spite of the discrepancies between the industrial practices and the academic research issues of concurrent engineering, concurrent engineering is a successful methodology for better product development from the industrial point of view. In this chapter, we try to analyze the problems of engineering design research community in contrast with concurrent engineering.

Engineering design research community traditionally worked on design theory and methodology. In my view, a design theory is an *abstract* system of theory to understand and describe design. This includes how to represent design objects and design processes, and therefore, a good design theory must serve as a guiding principle for developing CAD systems [13]. On the other hand, a design methodology is a system of theory (not necessarily at abstract level) to guide a design process of a specific class of machine or of artifacts in general. We can think of a concrete design methodology of spur gears, while there could be an abstract theory that targets technical systems design in general [14].

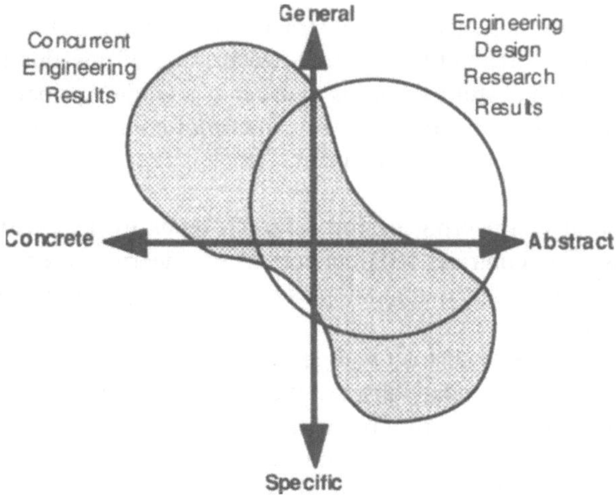

Fig. 2. *Categorization of research: Concurrent engineering mostly focuses on the "general and concrete" and "abstract and specific" categories, while engineering design research is interested in problems and theories "general and abstract."*

Here, we need to pay attention to the difference between "general" and "abstract." Despite the discrepancies that can be observed in industrial concurrent engineering practices and academic research issues, concurrent engineering addresses "general" problems, such as communication, organization, and technologies, but not "abstract" problems. This discussion leads to Figure 2 that categorizes theories in a two-dimensional space. One dimension is "general vs. specific," and the other is "abstract vs. concrete." Figure 2 compares research results of engineering design and concurrent engineering in this scheme.

(1) A specific and concrete theory is very useful for a particular application, but cannot be applied elsewhere. Its example is a design methodology of spur gears. Both concurrent engineering and engineering design research hardly address issues falling into this category.
(2) An example of the specific and abstract category is developing a computer program that works as an agent to solve a specific problem. Because computer programs deal with data, information, and knowledge of a specific domain, but not concrete entities, they are abstract. Some of both concurrent engineering and engineering design research eventually aim at developing computer software, hence falling into this category.
(3) Most of DfX methodologies fall into the general and concrete theory category. These theories can be applied to a variety of artifacts and are useful because the descriptions are concrete. Good DfX research results tell practitioners how to conduct a design process of a specific type of artifacts. Concurrent engineering is more interested in this category than engineering design research.

(4) The last category is general and abstract. As described before, such a theory can be extremely difficult to apply. However, one of possible applications of such a general and abstract theory is to use it as a guiding principle to develop a CAD system. This category is where concurrent engineering and engineering design research stands in sharp contrast.

A large body of engineering design research tends to deal with abstract and general theories. This contrasts with concurrent engineering research that resulted in general and concrete DfX methodologies or computer programs that are abstract but specific [15, 16].

Of course, any theory is abstract and general. Apparently, one of the missions of academic research is to generate abstract theories. The more abstract and general, the wider becomes the applicability of the theory, increasing the value of the theory. However, it would be extremely difficult to apply a too abstract theory. In contrast, concrete theories are useful, but because of their dependence on concrete problems, they cannot survive in rapidly evolving technology age. In fact, concurrent engineering research focused on general and concrete problems. If technology changes, the current concurrent engineering would not survive and we need to restart the research. A good example is demonstrated by Design for Disassemblability that basically made traditional modular design methodologies based on functionality obsolete. It completely revolutionized the ways to design modules for both better assemblability and disassemblability [17].

3.2. Lessons Learnt from Concurrent Engineering

Concurrent engineering aims at eliminating unnecessary changes and redesigns from a product development process, and at achieving better product quality. To do so, it is critical to facilitate teamwork and collaboration by improving communication among various participants in the development process and by allowing them to share information and knowledge. After reviewing concurrent engineering research and comparing it with engineering design research, we might draw the following lessons for the engineering design research community to improve and proceed its research.

(1) Focus on "general and concrete" problems. It might be helpful to work on a concrete test case and then to generalize it. This means that the engineering design research community should not emphasize research in "design methodology in general" but "design methodology for achieving better performance in some measurement." In other words, avoid "abstract" problems that cannot find immediate useful applications.

(2) Research should have such clear measurement of usefulness, though addressing usefulness too much is sometime dangerous. One measurement of usefulness might be innovativeness which is seldom addressed in concurrent engineering. Another measurement could be increasing design efficiency, but not productivity.

Concurrent engineering did so, by eliminating unnecessary changes and redesigns. Because practical design situations involve so many inefficient activities, it makes sense.

(3) Tackle design knowledge issues. Compared with concurrent engineering, there is very few research effort directly addressing "design knowledge" within the engineering design research community. We should analyze design knowledge in terms of design processes and design object representations, with emphasis on not "how" to represent but "what" to represent. Note that the "how" problem falls into the general and abstract category, as opposed to the "what" problem that falls into the specific and abstract category. This issue will be further discussed in the next section.

4. Knowledge as a Core Issue of Engineering Design Research

The world changes quite rapidly. The biggest and most influential change is perhaps in the principle of competition. Until the beginning of the 1990s, worldwide competition was simple; manufacturers of "more, cheaper, and better" products win. If we produce something more expensive with less quality, we sell less and eventually would be a loser [8]. Entering the mega-competition age, this situation changed. We must be able to achieve "quicker delivery of more innovative products." Customers simply want "quicker delivery" of better service. In addition, many consumer goods are sold in very much saturated, matured markets all over the world; faster production of more of the same may rapidly end up with being a great loser. This requests more added value in product's quality. Quality must not be just better but innovative. In terms of quantity, since the market is rapidly changing, faster production of more quantity does not suffice. We must be able to quickly responding.

From this point, we can signify the importance of knowledge as a core issue of engineering design research. Concurrent engineering is a methodology to work with more knowledge and information and, therefore, one of its core issues is sharing information and knowledge during product development. Beyond well-known communication, organizational, and technological problems which concurrent engineering addresses, we must focus more on knowledge as a source added-value. By intensively using product life cycle knowledge in an integrated manner, we can generate more added-value including innovation, longer life, higher reliability and robustness, more flexibility, and cheaper life cycle cost. To this end, we propose "knowledge intensive engineering" aiming at generation of more added-value through intensive and integrated use of knowledge, facilitating mutual communication among life cycle knowledge agents [18-20].

It is obvious through this discussion that "how" to represent such product life cycle knowledge is a minor issue. The major issue for the engineering design research community is "what" to represent. Traditional engineering design research focused on design theory and methodology and did not address these issues related to "what" to represent. Addressing "what" to represent belongs to the specific and abstract research category, which is a departure from the general and abstract research category that is very hard to find immediate applications.

The "what" problem consists of several issues. Concurrent engineering aims at facilitating mutual communication among participants in the product development process by sharing knowledge information. It is desirable, therefore, that the community should look at not only at computational frameworks for knowledge sharing but also product life cycle knowledge to be shared by those participants. Thus, the first of the "what" problem is the "ontology" problem. Ontology is a system of concepts that should be represented on a computer and it serves as a basis for knowledge sharing among different participants in the entire product life cycle. Without common ontology, achieving knowledge sharing is extremely difficult. In contrast, the CAD community has been working hard to establish a common product modeling scheme (e.g., STEP) emphasizing "how" to represent models of the design object rather than ontology.

Each of product life cycle should be supported by a "model." Therefore, the second issue is what kind of "model" is needed to represent a specific aspect in the product life cycle at an abstract level, so that every participant in the product life cycle can contribute better through model operations. So, the third issue is that "model operations" must be supported on a computational framework to facilitate better communication and to arrive at more added-value such as innovativeness. This issue also includes modeling of various activities including design. Notice that this is where research in design theory and methodology can play a crucial role in general and abstract level. Without sound design theory, it is hardly possible to formalize model operations.

A similar approach was taken by Cutkosky *et al.* [21] for the PACT project at Stanford University. However, even within the concurrent engineering community there is little research that addresses all of the ontology, model, and model operation issues in an integrated manner. Our group at the University of Tokyo has been working on a project to first develop "ontology" by systematizing product life cycle knowledge [22] This ontology serves as a basis for a "knowledge intensive engineering framework" for "knowledge intensive engineering" [18, 20].

5. Conclusions

This paper reviewed both industrial practices and academic research of concurrent engineering. While in general the manufacturing engineering research community concentrated in relatively narrow range topics, such as Design for Manufacture, the engineering design research community seemed to tackle a variety of issues related to concurrent engineering. So far, this strategy does not seem successful. In particular, in comparison with industrial practices, too much attention was paid to general and abstract problems, which resulted in theories difficult to apply. Instead, it is recommended that the design community should focus more on general and concrete problems and formulate methodologies that can improve design efficiency.

Further, one potential future research direction is "design knowledge" itself, not just how to represent it, but what to represent. This includes three issues. One is the ontology problem to build a knowledge sharing mechanism among various participants within a product life cycle. The second is the model issue to represent specific aspects in the product life cycle at an abstract level. The third issue is the model operation that should be supported on a knowledge sharing mechanism. As one of the conclusions of this paper, I suggest that the engineering research community should focus on these three as future core issues.

The history of CAD research can be characterized by epoch-making ideas introduced almost every 10 year. Sutherland coined the concept of "CAD" in 1963. The three-dimensional solid modeling technology bloomed around 1973. The knowledge-based CAD and intelligent CAD concepts began around 1983. What happened, then, in 1993? While it might be too early to conclude any at this moment, this knowledge issue related to concurrent engineering could be a candidate entry for this list of epoch-making ideas.

References

1. Ward A (ed) 1995 *Proceedings of the 7th International Conference on Design Theory and Methodology*. DE-Vol. 83, ASME.
2. Birmingham W P, Ward A 1995 Special issue: What is concurrent engineering. *Artif Intell Eng Des Anal Manuf* 9:67-68
3. Sriram D, Logcher R, Fukuda F (eds) 1991 *Computer-Aided Cooperative Product Development*. Springer-Verlag, Berlin (Lecture Notes in Computer Science No. 492)
4. Kusiak A (ed) 1992 *Concurrent Engineering: Automation, Tools, and Techniques*. John Wiley & Sons, New York
5. Sohlenius G 1992 Concurrent engineering. *Annals CIRP*, 41/2:645-655
6. Birmingham W.P, Ward A (eds) 1995 Special Issue on Concurrent Engineering, *Artif Intell Eng Des Anal Manuf* 9(2)
7. Dertouzos M L, Solow R M, Lester R K 1989 *Made in America*. The MIT Press, Cambridge, MA
8. Tomiyama T 1995 A Japanese view on concurrent engineering. *Artif Intell Eng Des Anal Manuf* 9:69-71
9. Ward A, Liker J, Sobek D, Cristiano J 1995 Toyota, concurrent engineering, and set-based design. In: Liker, Ettlie, Campbell (eds) *Engineering in Japan, Japanese Technology Management Practices*. Oxford University Press, Oxford, UK
10. Ishii K 1992 Modeling of concurrent engineering design. In: Kusiak A (ed) *Concurrent Engineering: Automation, Tools, and Techniques*. John Wiley & Sons, New York, pp 19-39
11. Finger S, Konda S, Subrahmanian E 1995 Concurrent design happens at the interfaces. *Artif Intell Eng Des Anal Manuf* 9:89-99
12. Boothroyd G, Dewhurst P 1983 *Design for Assembly: A Designer's Handbook*.

Boothroyd Dewhurst Inc., Wakerfield, RI.
13. Yoshikawa H 1981 General design theory and a CAD system. In: Sata T, Warman E A (eds): *Man-Machine Communication in CAD/CAM*. North-Holland, Amsterdam, pp 35-58
14. Hubka V 1987 *Principles of Engineering Design*, WDK 1. Heurista, Zürich
15. Finger S, Dixon J R 1989 A review of research in mechanical engineering design. Part I: Descriptive, prescriptive and computer-based models of design processes. *Res in Eng Des* 1:51-67
16. Finger S, Dixon J R 1989 A review of research in mechanical engineering design. Part II: Representations, analysis, and design for the life cycle. *Res in Eng Des* 1:121-137
17. Newcomb P, Bras B, Rosen D W 1996 Implications of modularity on product design for the life cycle. In: McCarthy J M (ed) *Proceedings of the 1996 ASME Design Engineering Technical Conferences and Computers in Engineering Conference*. 96-DETC/DTM-1516. ASME
18. Tomiyama T 1994 From general design theory to knowledge-intensive engineering. *Artif Intell Eng Des Anal Manuf* 8:319-333
19. Tomiyama T, Sakao T, Umeda Y, Baba Y 1995 The post-mass production paradigm, knowledge intensive engineering, and soft machines. In: Krause F-L, Jansen H (eds) *Life Cycle Modelling for Innovative Products and Processes*. Chapman & Hall, London, pp 369-380
20. Tomiyama T, Umeda Y, Ishii M, Yoshioka M, Kiriyama T 1996 Knowledge systematization for a knowledge intensive engineering framework. In: Tomiyama T, Mäntylä M, Finger S (eds) *Knowledge Intensive CAD, Vol 1*. Chapman & Hall, London, pp 33-52
21. Cutkosky M, Engelmore R, Fikes R, Genesereth M, Gruber T, Mark W, Tenenbaum J, Weber J 1993 PACT: An experiment in integrating concurrent engineering systems. *IEEE Comp* 26:28-37
22. Tomiyama T, Xue D, Umeda Y, Takeda H, Kiriyama T, Yoshikawa H 1992 Systematizing design knowledge for intelligent CAD systems. In: Olling G J, Kimura F (eds) *Human Aspects in Computer Integrated Manufacturing*, IFIP Transactions B-3. North-Holland, Amsterdam, pp 237-248

Understanding the Concurrent Engineering Implementation Process - A Study Using Focus Groups

Dr Fiona Lettice, Dr Stephen Evans and Palminder Smart
The CIM Institute, Cranfield University, Cranfield, Beds, MK43 0AL

Focus groups have been used as one method of collecting data on how practitioners view Concurrent Engineering and its implementation. Focus groups are a controversial research method, but their use provides a rich source of primary data, which can be supplemented by other data collection methods. The output from the focus groups shows the issues which Concurrent Engineering practitioners consider to be important for Concurrent Engineering implementation: senior management commitment, implementation planning, launching a multi-functional pilot team fast, and having a continuous improvement process. The need for extensive organisational analysis prior to implementation, for using team profiling techniques, early development of a new product strategy and investment in new technology have been understated by focus group participants.

1 Introduction

There has been a varied pattern of Concurrent Engineering (CE) implementation in the United Kingdom [1]. Although the value of Concurrent Engineering is generally recognised [2, 3] and the number of companies successfully implementing Concurrent Engineering is increasing [for example, 4, 5, 6, 7, 8, 9], there is still relatively limited penetration and understanding of the approach [9, 10].

The research, part of which is described in this paper, has the aim to develop an implementation methodology that companies can use to increase the benefits generated by Concurrent Engineering while reducing implementation costs, risks and time. The initial part of the research was primarily exploratory in nature, to determine the managerial and organisational factors important for successful Concurrent Engineering implementation. A qualitative data collection approach, the focus group method, was selected to provide a rich source of primary data. The use of focus groups, whose membership comprises Concurrent Engineering practitioners, has been supported by other methods, including literature analysis and in depth case studies [presented in 11, 12, 13].

The purpose of this paper is to show how focus groups have been used to identify the key factors for successful Concurrent Engineering implementation. Focus groups are a controversial research method [14] with many advantages and disadvantages, which are highlighted. The way focus groups have been used to collect data for this research is described and the main findings reported.

2 Focus Groups As A Research Method

Focus groups emerged as a research method to meet the need for the researcher to take a less directive and dominating role than is usually the case with standard interview techniques. The social scientists who developed and used the method felt that the respondent should be given more freedom to comment on what they perceived to be important. A focus group can be defined as a carefully planned discussion, involving between six and twelve participants, designed to obtain perceptions on a defined area of interest in a permissive, non-threatening environment [15]. Focus groups have been described as having five features: (a) people, who (b) possess certain characteristics, (c) provide data (d) of a qualitative nature (e) in a focused discussion [15].

Focus groups are not without their disadvantages [14, 15, 16, 17]. The groups can be difficult to assemble, as participants are asked to give up several hours or a whole day and travel to attend the session. The group tends to suffer from 'volunteer bias,' with a certain type of personality preferring to attend. The extra freedom given to the participants means that the researcher has less control over the direction of the discussion and emphasises the need for good facilitation skills. There is a danger that the participants may influence one another by interacting, or that the group is dominated by the most talkative members, causing weaker members to modify or even reverse their position on an issue. The researcher must be careful when analysing the findings, as it is important to interpret responses in their context and not to put too much emphasis on any one comment. The method is often criticised for not using a representative sample, making generalisation to the rest of the population difficult. The changing composition of the group also means that the results obtained are not replicable and it is hard to analyse the data collected and validate any conclusions due to the volume of data produced and the qualitative style of the method.

However, focus groups are recognised as providing a number of advantages over standard interviewing and questionnaire techniques [14, 15, 16, 17]. They provide a more natural setting for participants and the group setting is viewed as being more exciting and stimulating. The security of the group encourages its members to open up more, provokes spontaneity and generates a lot of new ideas. The freedom given to the participants allows them to answer in their own language and terminology, and limits the extent to which their answers are guided or restricted by a framework of questions or pre-coded answers. From the researcher's perspective, focus groups provide a rich source of data, from an acceptable sample size, while being less costly and time consuming to administer than interviews. The method gives the researcher the opportunity to understand the participants' viewpoint and problems and allows unanticipated issues to be easily explored.

3 Research Design

3.1 Requirements of Research Design

Initially, it was intended to conduct a telephone survey of Concurrent Engineering practitioners. The sample was selected to provide respondents from senior management positions with a strong involvement in CE and its implementation within their companies. They are the people in their organisation who are responsible for making change happen. They were also selected to have a broad understanding of the impact that CE will have on the whole business and not just the function they represent. Their names were found in articles about CE in engineering and business journals. The pilot telephone survey of twenty managers[1] reflected a high interest in the subject area, and led to the idea to use focus groups.

The focus group method was chosen to meet specific requirements:

- To benefit practitioners/participants by enabling them to share ideas and experiences
- To act as a data collection exercise for research
- To raise the profile of The CIM Institute, Cranfield University as an information platform for Concurrent Engineering

3.2 Participants

The focus groups have been held two to three times a year over a four year period. At the first focus group meetings, the participants were predominantly from technical and engineering functions and approximately 50% of those contacted in the initial telephone survey participated. The group has evolved to include

[1] It should be noted that at the time of the survey (1991), CE was not well understood or widely practised. This made it difficult to find many companies who had successfully implemented CE.

representatives from non-technical functions. New members are usually recruited by existing participants. The companies represented are all from the manufacturing sector, including the white goods (2 companies), automotive (3 companies), aerospace and suppliers (5 companies) and electronics (5 companies) industries. The companies involved in the focus group vary considerably in organisation size (from about 300 to over 3000 employees), product type, manufacturing processes, and product development strategy. They do however have one thing in common; they have all successfully implemented Concurrent Engineering, using multi-disciplinary product development teams, to achieve 30 to 50% improvements in time to market, quality and cost related metrics. Each company has different experience with CE methods, tools and techniques, which they have adapted and used as the need arose. The companies are at different stages in their CE implementation, bringing a range of experience and knowledge of CE to the focus group. All of the participants have been and are directly involved in introducing and using Concurrent Engineering principles in their organisations; some for more than a decade, while others have only been practising CE for three or four years. Two of the participants have experience of implementing Concurrent Engineering in more than one company.

3.3 Focus Group Meetings

The agenda for each day begins with an introductory ice breaker exercise. To give the participants time to explore and discuss the issues raised, only two or three questions can be asked. Open ended questions are used, so that the participants can determine the direction of the response. The questions are designed, by the researcher, to evoke group discussion. "Why" questions are avoided or replaced with "What" and "How" questions, which are less directive. In some cases a short presentation may be given to set the context of the question. The questions are not set with the aim of arriving at a solution, but for exploring the various dimensions of an issue. Examples of questions that have been used in the CE focus groups are shown in Figure 1. These questions all probe the same issue. By asking the same question in a number of different ways, a wider range of responses can be obtained.

What are the main obstacles to implementing Concurrent Engineering?

How to create the optimal organisation structure - what are the key characteristics that allow an organisation to successfully implement CE?

How would you know a good CE implementation when you see one?

How would you accelerate/improve the implementation of CE?

Fig. 1. *Examples of Concurrent Engineering Focus Group Questions*

The participants are split randomly into two mini groups of three to five members to answer the questions. A technique called "Creative Silence" [18] is used to start the process of brainstorming. The participants individually generate ideas and solutions in response to the question posed. The participants are asked to use post-its, and write one idea per post-it. Each participant then contributes their ideas singly and without comment from the group until all of the ideas are represented. This approach prevents the first idea presented from being seized upon, allowing a greater diversity of ideas to be generated. It also means that weaker members of the group have an equal and early opportunity to contribute their ideas. Participants are encouraged to listen to one another, and only speak if a contribution needs clarifying. After the initial ideas are collected, further generation of ideas may occur. The groups are asked to collect and move their ideas into families on a white board or flipchart. At this point gaps may be identified and filled by another wave of Creative Silence. At the end of the allocated time period, the groups are asked to present their outputs to one another. Time is allocated for a discussion to follow, comparing the output from the two groups and recounting personal experiences. There is no pressure to reach a consensus or arrive at a "solution."

For subsequent exercises during the day, the original groups are again randomly broken up and new groups formed, to encourage a mixture of ideas and personalities and to prevent more talkative participants from dominating the whole day.

At the end of the day, a review exercise is carried out, using the same Creative Silence brainstorming technique. The aim is to continually refine the focus group process and to allow the participants to be more actively involved with running the focus groups. A high proportion of the participants have attended nearly all of the focus group meetings, providing a high level of continuity over time, and increasing the confidence of the group to tackle more complex issues. The participants are invited to contribute topics they would like to see covered in subsequent group meetings. If the participants face particular problems in their CE implementations, they have the opportunity to give a short presentation of their problem to the focus group. This technique minimises unintentional bias from the researcher and shifts attention towards the participants.

A moderator or facilitator, who is comfortable and familiar with group processes and dynamics, hosts the day. Their responsibilities include maintaining enthusiasm during group discussions, lightly guiding discussion when necessary, and time-keeping. The moderator is supported by one or two assistant moderators whose role is to take notes, respond to unexpected interruptions, keep refreshments flowing and respond to participant needs for facilities during the day. After the meeting, the moderators perform a post meeting analysis to ensure that the main points have been captured and to discuss organisational aspects of the focus group activity. Originally the facilitator role was performed by a researcher, but as the focus group has matured, occasionally this role has been assumed by a participant volunteer. This helps to further remove unintentional researcher bias and demonstrates a robustness of process. All of the output from the day is typed up and circulated to participants for personal use and future reference. Care is taken to present the data as it appeared during the meeting, without the researchers adding their interpretation to it.

4 Discussion Of Results

4.1 Definition of Concurrent Engineering

During the first focus group meeting held, the group were asked to agree on a definition, or definitions, of Concurrent Engineering. The aim was to understand each others views of what CE is, explore the differences and similarities, and finally combine the different viewpoints to produce a single definition that was acceptable to all participants. This definition would then form a common reference point for future focus group discussions and activities. The definition arrived at was:

Concurrent Engineering is the delivery of better, cheaper, faster products to market by a lean way of working, using multi-discipline teams, right first time methods and parallel processing activities to continuously consider all constraints.

A lean way of working means removing non-value adding activities from the product development process and stresses not adding extra resources overall to achieve shorter product development cycles

Multi-discipline teams are the essential basic building blocks of CE which bring together members from all relevant functions involved throughout the product development project life cycle. Members are selected to achieve the right skills mix. Any skills lacking are identified early and can be contracted into the team when they are needed.

Right first time methods cover a variety of techniques and technologies that help to consider multiple constraints concurrently and typically earlier than previously (for example, Design for Assembly, Quality Function Deployment and Early Supplier Involvement); and store, manipulate and communicate design information (for example, Computer Aided Design and Engineering Data Management systems).

Parallel processing activities is, where possible, beginning the next task before the previous one is 100% complete. This involves early sharing of incomplete information between different activities in the product development process.

To continuously consider all constraints means appreciating the impact of early design decisions on for example the manufacture, maintenance and reliability of the product. The aim is to maximise the number of constraints considered at the beginning of the product development process.

The discussion generated while developing the definition centred around two main points. The first was that the term Concurrent Engineering is misleading. It implies a smaller scope of change than is actually the case. The CE practitioners concurred that Concurrent Engineering should be viewed as a broad organisational change, rather than one limited to the design and manufacturing engineering functions. The second point was related as they believe that when improving the product development process, the engineering issues are small, whereas the human resources or people issues are paramount. This means that in many ways, CE is not so different from other organisational changes involving team working and employee empowerment. Where CE is different is that it has a strong focus on improving primarily the process of product development, and removing functional boundaries between all of the departments involved in product development. Most often, emphasis is placed on removing the wall between design and manufacturing, but the walls between design, marketing, purchasing, customer service and so on are equally important.

Subsequent focus group meetings explored Concurrent Engineering and its implementation in more detail. The responses to three of the questions are shown and their significance discussed under two main headings: process of implementation and approach to implementation. When given the question "What are the key stages to implementing Concurrent Engineering?," one group came up with the output shown in Figure 2. Another question used was "What are the obstacles to implementing Concurrent Engineering?". The groups brainstormed their ideas and then grouped their post-its into families. One group used an Ishikawa diagram to display their output, which is copied directly from the material produced and shown in Figure 3. A similar question, "How would you accelerate/improve the implementation of Concurrent Engineering?," produced the output shown in Figure 4.

Acceptance	Will to Change/Accept Change is Needed Sponsorship Upgraded
Focusing	What Do We Need to Do? Route (Now => Then)
Preparing	Ownership Opening Up Enrolling ===> Pilot Empowerment/Support Selection of Team
Learning (Implementing)	Pilot - Short/Sharp Results Learning and Appraisal Methods (Structure, Management, Training, Impact)
Continuous Improvement	Roll Out (Remove Barriers, Extend) Institutionalise Processes etc. Common Processes

Fig. 2. *Output to Question "What are the key stages to implementing CE?"*

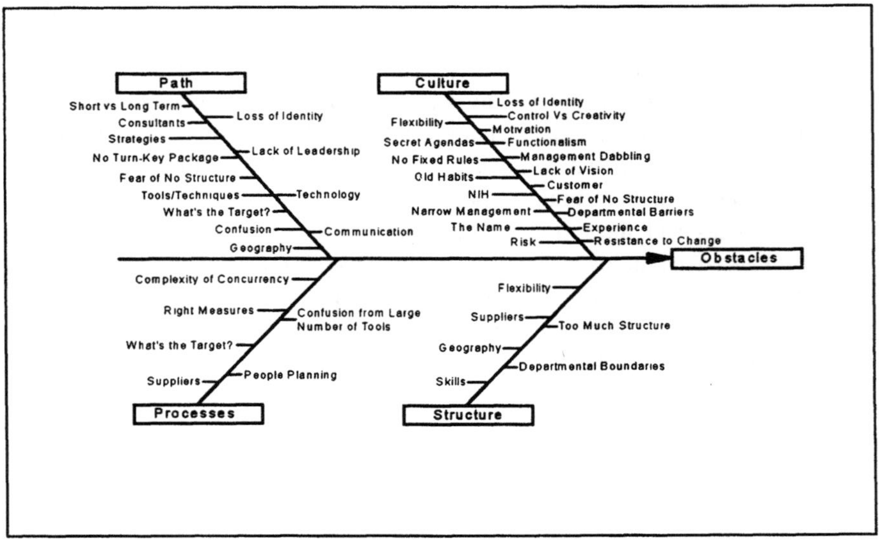

Fig. 3. *Output to Question: "What are the obstacles to implementing CE?"*

4.2 Process of Implementation

By analysing the outputs, a number of observations can be made. Concurrent Engineering practitioners have identified management obstacles to change which include a lack of leadership, a lack of vision, a lack of motivation, and a tendency to take a short-term rather than a long-term view. Overcoming these obstacles requires top management commitment to the CE implementation, a view supported by many [for example, 19, 20, 21, 22]. CE must be defined as a strategic direction to be taken and coupled closely with the overall business strategy. All too often CE is treated as an engineering initiative, and is only supported by the engineering and technical departments. To achieve radical improvements in product development performance requires a truly cross functional approach, which can only be sanctioned by those at the top of the organisation. Improvements in the region of 20-60% are not made through localised decisions, but through the involvement, co-operation and understanding of many other departments. Even seemingly simple product development decisions will impact upon, for example, purchasing strategy, manufacturing strategy, marketing strategy and the accounting and computing systems used. High level sponsorship, often demonstrated through the formation of a Steering Group of senior managers, signals a high degree of commitment and provides a broad and strategic view over the many functions involved. Management must find clear ways to show they are interested and willing to support improvements to the product development process. This is usually achieved by devoting significant proportions of their time to what the CE practitioners have termed the "selling job."

Using a pilot project team, with members from many functions, is recognised as an important method for building up momentum to change. Where possible, key suppliers and customers should be included on the team. Practitioners believe that the cross-functional team is the most important element of CE, and that the emphasis should lie with setting up a good team environment, before considering which of the many CE tools and technologies to implement. Ideally the team should be collocated. Companies who gave the team their own dedicated work area or office later in the product development cycle consistently wish that they had done it sooner, in order to rapidly improve communication between team members and build a coherent team identity. The pilot project is a fast way to learn and build in-house CE expertise, and provides a positive step towards breaking down departmental barriers. If the team is to be successful, then it should be empowered to make product development decisions, set its own direction and goals within management guidelines [23] and change and develop the product development process. Many of the focus group participants advocate selecting a pilot project with a high likelihood of being successful to improve morale and help spread the new way of working more quickly.

Drivers
Top level steering group of on- and off-line management
Be prepared to take autocratic action - recognise criteria for this first
Use external consultants (?)
Map out key stages - responsibilities, objectives
Top down and bottom up approach
Involve senior management in making legitimate the objectives

The Selling Job
Identify advantages of CE to all
Identify advantages for business and individual
Increased profit, job security, job enrichment

Analysis and Evaluation
Identify key processes and measure current performance
Be prepared to re-evaluate performance measurement process

Set Standards
Investigate and observe best practices
Benchmark/collaborate with others in a structured way
Set up Critical Success Factors for pieces of work - do not start if all not present

Training
Train leaders and facilitators
Communicate opportunities of CE
Select people to lead process

Demonstrator
Choose pilot piece of work through process - put team around it
Pilot representativeness (i.e. understand context)
Allow team to develop new process ==> roll-out
Genuine recognition for SE behaviour successes
Provide feedback in a structured way
Publicise successes at team and senior level

Roll-out/Scale Up
Use new process measures
Staged reviews
All new work (past pilot) to enter organisation by CE principles

Fig. 4. *Output to Question: "How would you accelerate/improve the implementation of CE?"*

The pilot project can be used as a company demonstrator, to publicise changes and win over sceptics at all levels in the organisation. Areas of highest resistance can be ignored in the early stages of implementation, returning to them later when sufficient momentum has been attained. The pilot team can also be extremely flexible, and provide an experimental test bed for new ideas and approaches.

The pilot team leads naturally into the next phase of implementation, where the problem of expanding the CE methods and mode of working into subsequent product development projects must be tackled. Team members from the first pilot project can be migrated to subsequent projects to facilitate the rapid adoption of new ways of working and developing cheaper, faster and better products. It is at this point that senior managers can strategically plan the implementation, with the understanding gained from the pilot project. The right tools and technologies can be identified for the unique product development problems faced. Training needs can also be identified, which will range from team building and simple problem solving skills through to detailed technical and product-related training. The main aim of the "roll-out" phase is continuous improvement: building on previous successes and learning from previous mistakes. Key individuals are assigned to subsequent projects to facilitate this [23]. New processes and behaviours, which are not project-specific, can be institutionalised by continually applying them and properly supporting them until they become the norm.

4.3 Approach to Implementation

One of the recurring messages from the focus group meetings is that the CE pilot team should be launched as soon as possible. Concurrent Engineering requires change on a broad front, but all of the elements required do not have to line up right from the start. By starting small, a degree of comfort can be generated and the risk of making costly mistakes can be minimised. Effort should be placed in selecting the right pilot project. This should be an important project to the company, for example the introduction of a new product rather than smaller, less complex projects requiring only incremental improvements or enhancements to existing products. The team should be provided with a clear product specification, product development target and distinct boundaries for making product development decisions. The pilot team should then begin product development activities as soon as possible. This encourages the company to learn fast through experience. A survey into UK product development [24] revealed that "some 60% of CE expertise is gained through self-learning." This fast action approach to CE implementation differs from some popular approaches.

CE practitioners believe that a thorough analysis of the business is not required before implementation can begin. Ideally everyone involved should understand the principles and language of CE, but more importantly, a common sense approach is required. The temptation to over-theorise should be avoided. This contradicts the approach advocated by Karandikar [25] and Mentor Graphics [26]. They classify the attributes of Concurrent Engineering into categories. A company then uses the classification system to quantitatively analyse the current state of their product development process and infrastructure and so assess their readiness to implement CE. The readiness tools help them to identify which elements of CE to implement first. The opinion of mature CE practitioners suggests that this approach tends to lead to "paralysis by analysis," where companies become obsessed with finding out more and more about how they currently develop products and how others use Concurrent Engineering, rather than experimenting with new ways to develop products and improve processes. The readiness assessment tools are useful later in the CE implementation programme, when the teams and management are struggling to find ways to continuously improve. They also provide useful educational aids, showing how the various elements of CE fit together and where emphasis can be placed to achieve different improvement goals.

The focus group output also shows a strong opinion that the team members should not be selected using personality profiles and tests, such as those developed by Belbin [27] and Myers-Brigg [28]. Team members should be selected to represent a function or skill that the product development project requires. Team members can initially be selected on their keenness to be involved, the technical competencies they can bring, and their ability to work flexibly with other people in a team environment [29]. These skills require no analysis as they are obvious to managers and colleagues alike. A company often will not have the luxury to choose the most ideal personality profiles for team working, and should not be deterred from using product development teams for this reason. More mature teams may wish to use team profiling tools to identify their strengths and weaknesses, but in the early stages they provide another level of complexity to slow down implementation progress.

The CE practitioners do not consider strategy planning to be a major issue in the early phases of implementation. Although in the long term it is beneficial for the business to have a well-defined new product development strategy [30], which fits the business strategy, this is not a pre-requisite to begin CE implementation. A new product development strategy must be developed by senior management and can be developed alongside the pilot team activities, to aid the selection of the next few projects. The temptation to delay implementation until a new product strategy is developed should be avoided. A new product development strategy will help to set clear product objectives [31] and greatly reduce confusion at product development review meetings. The first review meetings of the pilot team will naturally highlight and drive the need to develop a better strategy [32, 33].

The role of technology in CE implementation is de-emphasised by the CE practitioners. They feel that the technology and its implementation take over and the issues surrounding setting up a supportive team environment can easily get overlooked. During the focus group meetings, CE tools and technologies and their implementation and use are rarely mentioned, with the participants preferring to focus on what they consider to be the most important aspects of CE: people and the effective implementation of empowered product development teams. There must be an emphasis on computer technology *and* face-to-face communication, rather than a substitution of one for the other [34]. Tucker and Leonard support this view based on their experience of CE implementation, stating that "CE is fundamentally a people driven concept, not technological" [35]. Again, technology needs will arise and should be implemented to squeeze maximum efficiency from an already improved and streamlined process. Early automation, without understanding the new process fully can lead to the same mistakes being made faster and at a greater cost to the company.

5 Conclusions

The output from the focus groups has been complemented by a thorough review of the literature and a questionnaire and interview survey of companies who have implemented Concurrent Engineering [13]. From this knowledge of what works, a methodology for implementing Concurrent Engineering has been developed. This has been delivered and tested through live application in collaborating companies in the discrete manufacturing industry. The development and structure of the methodology is described elsewhere [7, 12, 36]. The output from the focus group research method is also being used as the starting point for further investigation into the details of best practice Concurrent Engineering implementation. This is being conducted via semi-structured interviews and questionnaires in a world-wide survey of companies, at various stages in their Concurrent Engineering implementation.

This paper has shown how focus groups can be used to collect data from Concurrent Engineering practitioners. The advantages of the method are that the participants generate considerable data and ideas, using their own language and terminology. From the researcher's viewpoint, the method has been less costly and time consuming to administer than interviews. The main disadvantage of the method has been the volume of qualitative data produced and the inherent problems of analysing it and interpreting responses in their context with the appropriate level of emphasis. The data is therefore being supported, in the broader research project, with quantitative and qualitative data from other sources to improve validity. The focus groups have been used over a four year period, with participants from various manufacturing companies who have implemented and practice Concurrent Engineering.

Many of the questions used have focused on the successful implementation of Concurrent Engineering. The output has highlighted the critical nature of managerial and organisational issues, in particular the need for senior management commitment, implementation planning, launching a multi-functional pilot team fast, and having a continuous improvement process. The focus group participants cannot stress strongly enough the vital role that 'learning by doing' plays in the rapid implementation of CE and improved product development performance. The need for extensive organisational analysis prior to implementation, for having the correct team personalities, early new product strategy development and investment in new technology have been consistently understated by participants in this research.

Acknowledgements

This work forms part of the research project, ***FAST CE - Implementation Methodology for Concurrent Engineering*** (GR/J/57735), which is jointly sponsored by Computervision UK Ltd and the Control, Design and Production (CDP) Group of the Engineering and Physical Sciences Research Council (EPSRC) of Great Britain.

References

[1] Coughlan P D, Voss C A 1992 *Review of Practice and Issues in Simultaneous Engineering in the UK*, London Business School, July (supported by SERC Grant References GR/G62400; GR/G63605)

[2] Belson D and Nickelson D 1992 Measuring the economics of concurrent engineering. In *CE & CALS Conference and Exposition*, pp 539-549

[3] Hartley J and Mortimer J 1990 *Simultaneous Engineering: The Management Guide,* Industrial Newsletters Ltd, Dunstable

[4] Barton R J 1994 Concurrent Development of Radar Systems. In *IEE Colloquium on Current Developments in Concurrent Engineering - Methodologies and Tools*, (Digest Number 1994/140)

[5] Computervision 1994 Concurrent Engineering: Winning the argument is just the beginning. *Computer Aided Draughting and Design*, pp 12-13

[6] Ellis D 1992 Success at the drawing board. *Infomatics*, March, pp 47-54

[7] Evans S, Lettice F E and Smart P 1994 Concurrent Engineering - Key Implementation Issues. In: Kidd P T, Karwowski W (eds) *Advances in Agile Manufacturing*, IOS Press

[8] Manton S M 1991 Concurrent Engineering for the Land Rover Discovery. In: *CALS Europe 91 Conference Proceedings, Part 2*, pp 593-598

[9] Nichols K 1994 Developing with the Best: Survey of product development within UK industry. *World Class Design to Manufacture*, 1(2): 7-12

[10] Benchmark Research 1994 *The 1994 Manufacturing Attitudes Survey - Management Summary*, Initiated by Computervision, Coventry

[11] Bowden W and Lettice F 1996 Changing the Organisation Through People at Measurement Technology Ltd. In: Backhouse C J, Brookes N J (eds)

Concurrent Engineering: What's Working Where, Gower/Design Council, Aldershot, pp 149-162

[12] Lettice F E 1995 Concurrent Engineering: A Team-Based Approach to Rapid Implementation, PhD Thesis, The CIM Institute, Cranfield University

[13] Smart P K, Lettice F E and Evans S 1996 A UK Based Empirical Investigation of the Factors Contributing to a Successful Implementation of Concurrent Engineering. In:, *Managing Integrated Manufacturing (MIM) Second International Conference*, 26-28 June

[14] Drayton J L, Fahad G A and Tynan A C 1989 The focus group: a controversial research technique. *Graduate Management Research,* pp 34-51

[15] Krueger R A 1988 *Focus Groups: A Practical Guide for Applied Research.* Sage Publications, Newbury Park, CA

[16] Hoinville G and Jowell R 1978 *Survey Research Practice,* Heinemann Educational Books, London

[17] Tull D S and Hawkins D I 1987 *Marketing Research: Measurement and Method.* MacMillan Publishing Company, New York

[18] Newman V 1994 *The Creative Manager: A Toolbox for Problem Solvers,* Cranfield University: The CIM Institute.

[19] Byrd J and Wood R T 1991 Implementation Strategies for Integrated Product Development In: *CALS Europe 91 Conference Proceedings, Part 2*, pp 549-560

[20] Knodle M S 1991 *Transitioning to a Concurrent Engineering environment,* American Institute of Aeronautics and Astronautics, AIAA-91-3151

[21] Smith P G and Reinertsen D G 1991 *Developing Products in Half the Time,* Van Nostrand Reinhold, New York

[22] Zangwill W I 1993 *Lightning strategies for innovation: How the world's best firms create new products,* Lexington Books, MacMillan, New York

[23] Takeuchi H and Nonaka I 1986 The New New Product Development Game - Stop Running the Relay Race and Take Up Rugby. *Harvard Business Review,* January/February. pp 137-146

[24] Costanzo L 1993 Breaking out is hard to do. *Engineering,* April, pp 25-26

[25] Karandikar H M, Fotta M E, Lawson M, Wood R T 1993 Assessing Organisational Readiness for Implementing Concurrent Engineering Practices and Collaborative Technologies In: *Proceedings of the Second Workshop on Enabling Technologies: Infrastructure for Collaborative Enterprises,* Morgantown, WV, pp 83-93

[26] Carter D E and Baker B S 1992 *Concurrent Engineering: The product development environment for the 1990s,* Addison Wesley Publishing Company, Reading, MA

[27] Belbin R M 1981 *Management Teams: Why they succeed or fail,* Heinemann, London

[28] Charney C 1991 *Time to Market - Reducing Product Lead Time,* Society of Manufacturing Engineers, Dearborn, Michigan

[29] Garrett R W 1990 Eight steps to Simultaneous Engineering. *Manufacturing Engineering*, Nov, pp 41-47

[30] Booz, Allen and Hamilton 1982 *New products management for the 1980s.* Booz, Allen and Hamilton Inc, New York

[31] Twigg D and Voss C A 1992 *Managing Integration in CAD/CAM and Simultaneous Engineering - A Workbook,* Chapman and Hall, London

[32] McGrath M E, Anthony M T and Shapiro A R 1992 *Product Development: Success through Product and Cycle-Time Excellence*, Butterworth-Heinemann, Boston, MA

[33] Wheelwright S C and Clark K B 1992 *Revolutionizing Product Development: Quantum Leaps in Speed, Efficiency, and Quality*, The Free Press, MacMillan, New York

[34] Clark K B and Fujimoto T 1991 *Product Development Performance - Strategy, Organisation and Management in the World Auto Industry*, Harvard Business School Press, Boston, Massachusetts

[35] Tucker D E and Leonard R 1994 Overcoming the Cultural Barriers to Implementing Concurrent Engineering. In: Kidd P T and Karwowski W (eds) *Advances in Agile Manufacturing*, IOS Press

[36] Lettice F E, Smart P K and Evans S 1995 A Workbook-Based Methodology for implementing Concurrent Engineering. *International Journal of Industrial Ergonomics*, Vol 16, pp 339-351

Architecture to Handle Concurrent Engineering

Christophe Cointe, Nada Matta
INRIA (projet ACACIA)
2004 route des Lucioles BP. 93
06902 Sophia-Antipolis Cedex
e-mail: {ccointe,nmatta}@sophia.inria.fr

ABSTRACT
Concurrent Engineering is a hard task in which several designers collaborate together to build a system given requirements. To guide Concurrent Engineering task modelling, we define a generic model, in which individual designs are combined with coordination. Characteristics and nature of each sub-task are emphasized through this model. In order to guide Concurrent Engineering representation, we define a structure based on design statement representation. Each individual design and communicated information is represented as an alternative from which a solution (design statement with single alternative) is determined.

Keywords: Concurrent Engineering, Conflict, Knowledge Representation, Information Exchange, Task Modelling.

Introduction

In Concurrent Engineering (CE), several experts (designers, managers,... called participants in this paper) in different specialities, collaborate together in order to construct a system (also called artifact), given the customer's requirements. Each designer uses knowledge learned from his experience and also shared knowledge from his collaboration with the designer group. Conflicts can be revealed from disagreements between designers about proposed designs (called propositions) and needs [8]. Such conflicts are detected in CE and negotiation methods are used to solve them.

To handle CE, different types of knowledge will be distinguished and represented, e.g: Task organization, communication protocols, conflict management, knowledge sharing, task realization,...

Our main objective is to study CE task modelling and representation. We define generic models to help to model this task and specially the conflict management task. These models will provide generic components to be used (by a knowledge engineer[1]) as building blocks in CE and conflict management task modelling for a real application. We also determine a structure to support this task and to represent its particular components. This work is a part of the Genie[2] project, in which the CE task is analysed in order to provide guides for defining corporate memory.

We present in this paper a generic CE task model (I) in which individual propositions are combined with coordination. A structure to implement this model is then described (II), in which a structure is proposed to help propositions formulating and evaluating. Finally, we present how the structure can be used by designers to realize the CE task (III).

I. Model for Concurrent Engineering Task

The CE task model that we defined, is represented at knowledge level [10], in which knowledge characteristics and nature are described. There is also a distinction between this level and symbolic level in which implementation choices are made.

In order to define a generic CE task model, we have represented models, we studied from the literature, in a single formalism. This allows to analyse their characteristics and to reveal their respective contributions. The generic components that we defined are presented as an extension of the CommonKADS library [5]. So, they are described with the CML[3] [16], the conceptual language offered by CommonKADS to represent expertise models. The generic models, provided by CommonKADS library for tasks such as diagnosis, design, etc., are described in CML. Therefore, it seemed natural to exploit CML to describe a generic model of the CE task.

I. 1. Concurrent Engineering Task Modelling Research Overview

A number of researchers have studied CE tasks and provided models of this task, like Ramesh et al. [11], Sycara [15], Brazier et al. [4] and Bond [2]. Bond's and Brazier et al's models offer a global view upon the main particularities of CE. We abstract as follows the characteristics emphasized in these models.

In this section, figures represent only data flows like in CommonKADS Task structures [5]. Rectangles show tasks input/output, rectangles with rounded angles present tasks.

1.A Knowledge Engineer is able to acquire knowledge and define a model for real applications.

2.Genie project is a collaborative project between INRIA and DASSAULT AVIATION.

3.CML: (Conceptual Modelling Language) is defined in CommonKADS approach [Breuker et al,94].

a) In Bond's model, two important notions appeared: Shared model and Private models. Each private model belongs to a participant of the concurrent engineering task and describes his knowledge. In fact, a private model is the expertise model of a participant (i.e. an explicitation of the model of his knowledge). These expertise models are private and they are not shared. However, the Shared model describes a shared model of the system to be designed.

First some participants use their private models to propose modifications in the artifact. These modifications are then communicated to the other participants. If there is no conflict, the Shared model and possibly private models are modified, and so on (Figure 1).

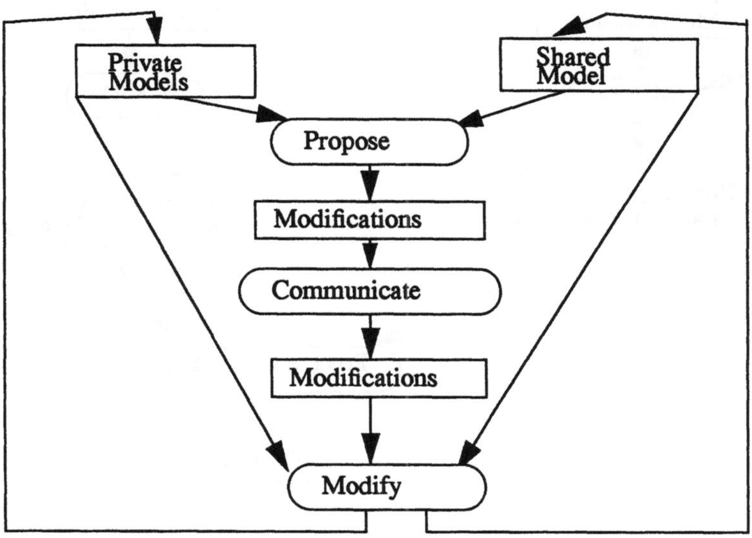

Figure 1. Bond's Concurrent Engineering Model.

b) In their model, Brazier et al. highlight the fact that in Concurrent Engineering, modifications can be made in requirements as well as in the artifact model (Figure 2). So, conflicts can be revealed and managed in requirements as well as in artifact model modifications.

A coordination task (Figure 2) determines whether requirements or the artifact model must be modified in the next design step, and which part of requirements or the artifact model. This task decides also if the design process can be continued or stopped (if a deadlock cannot be solved, for example).

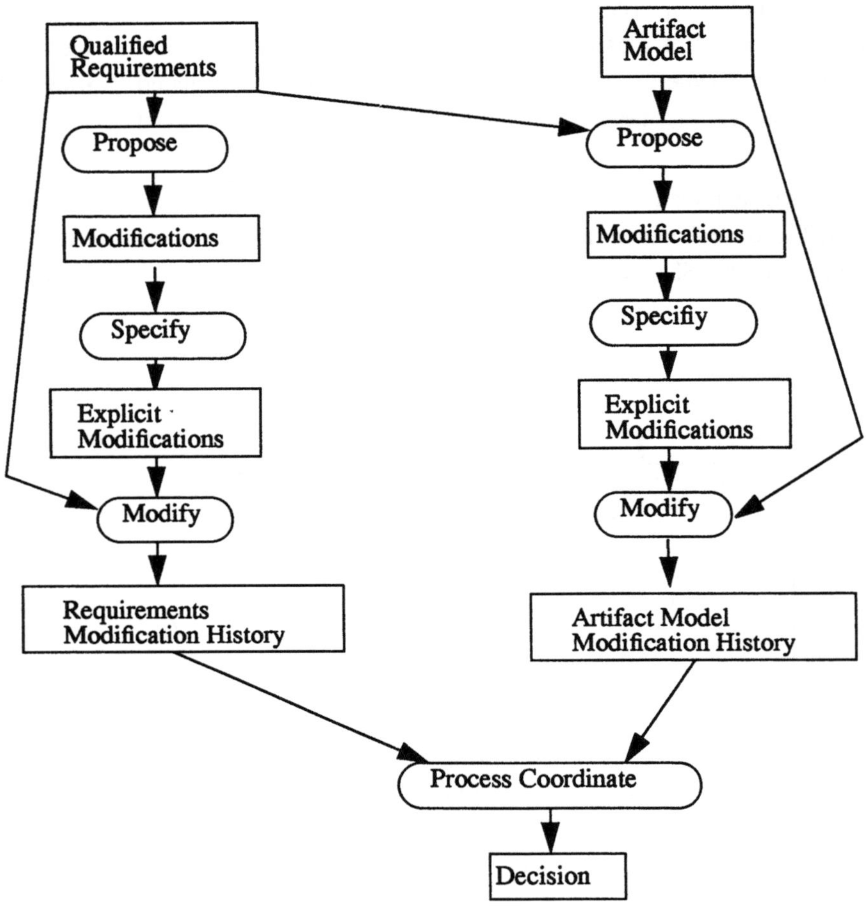

Figure 2. Brazier et al.'s Concurrent Engineering Model.

In each step in this process, conflicts can be detected. For example, a satisfaction conflict can appear when a participant proposes some modifications which do not satisfy requirements. Specifications of propositions may be also in conflict with other participants' objects description and needs. For more details see [4].

I. 2. Generic Model for Concurrent Engineering Task

After studying Concurrent engineering models, we propose a generic model for the concurrent engineering task, in which private propositions are combined with coordination (Figure3).

Three steps at three levels are performed in the CE task (Figure 3):

- Personal level («Design» task). At first, each designer generates some propositions to satisfy given requirements, based on his private knowledge (Private model). His task is identical to a design task. The generic design model represented in the CommonKADS library [5] can guide the modelling of this task.

- Inter-personal level («Argument» task). To communicate his propositions to the group, a participant arguments them. Assumptions made in the first step, are used to determine arguments.

- Group level («Evaluate» task). The last step is an evaluation task. Propositions may not satisfy participants' needs and conflicts can appear. So, the principal subtask in this evaluation consists of detecting and solving conflicts.

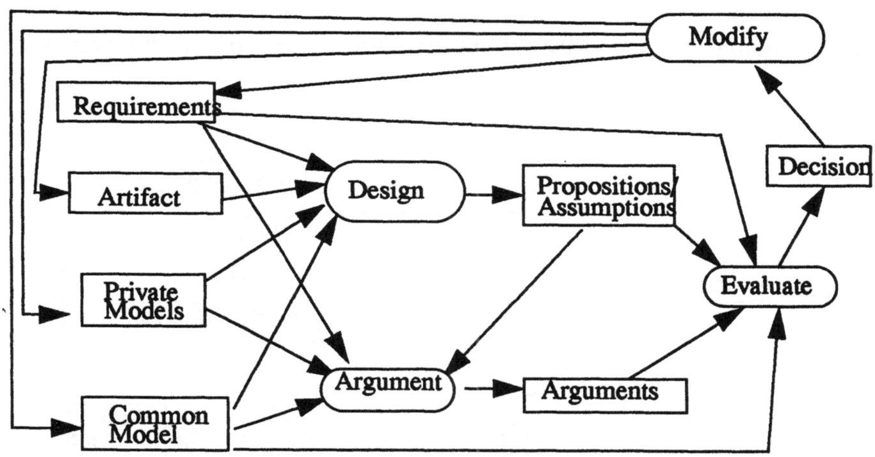

Figure 3. Generic Concurrent Engineering Model.

Propositions can concern the requirements or the artifact model. Decisions about which object (requirements or artifact model) can be modified in the next process cycle and in which part modifications can be made, are taken in the design control (a subtask of "Evaluate").

Sometimes, new knowledge can be learned from this experience. Then, some private models need to be modified by their respective owner. In the same manner, the Shared model can be reorganized and modified, because other shared knowledge can be explicited at each cycle.

In the next sections, we detail the tasks realized in the three levels.

I. 2. 1. Design Task

When a participant generates a proposition, his task is a design task. He must propose a design respecting related requirements. According to the design generic model proposed in CommonKADS library [5], a participant first specifies the problem, then generates the corresponding subpart of design and evaluates if this subpart satisfies given requirements. Violated requirements cause modifications in the generated subpart of design or in the requirements.

In a CE task, a participant needs also to extract propositions which are more understandable by other participants than the corresponding subpart of design (Figure 4.). These propositions of design are then communicated to all the group of participants.

In CE, a Shared model is needed to facilitate cooperation between participants. The explicitation of shared knowledge enables a global view of the system on the one hand, and helps to determine interdependencies between participants on the other hand. The Artifact represents the current state of the system to be built (Figure 4.).

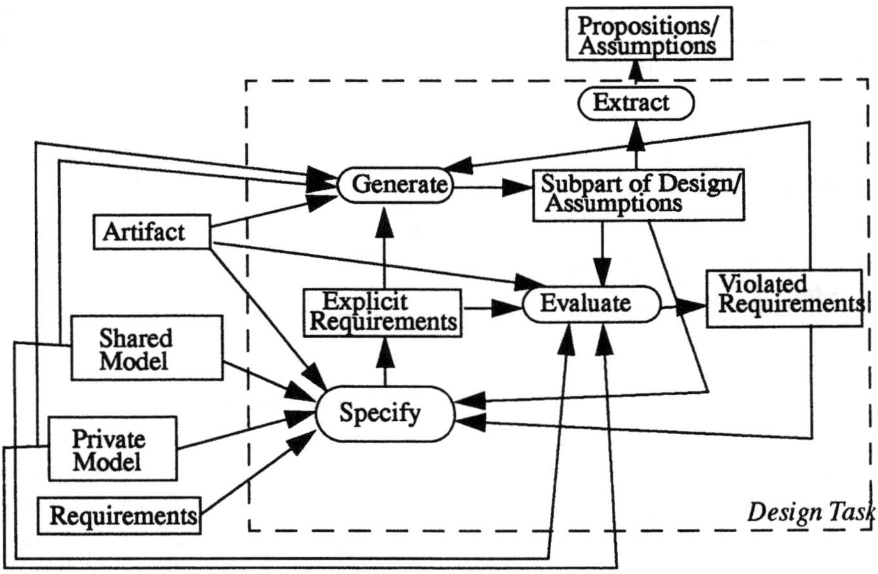

Figure 4. The «Design» task Model.

I. 2. 2. Argument Task

Argumentation is used in CE to defend a proposition and to persuade the group to choose it. Arguments must be explicit either to avoid conflicts if possible or to determine precisely their nature. Propositions are first justified with potential arguments which are then communicated to the group with the corresponding propositions. These arguments may not be sufficient to force the choice of propositions at the next level «Group level» and some of the arguments may be refuted by the group. So, a backtrack can be made and potential arguments are evaluated considering eventual conflict and refuted arguments (Figure 5).

A number of methods are proposed to help a participant to argument his proposition. K. Sycara [14] recommends Case-Based Reasoning (CBR) to define arguments by retrieving them from previous cases. She proposes also some strategies using general principles and heuristic rules to define arguments. B. Ramesh [11] favours interdependency sharing to detect potential differences in order to consider them in the definition of the arguments.

I. 2. 3. Evaluate Task

Propositions may not satisfy participants' needs and conflicts can appear. So, the principal subtask in this evaluation consists of detecting and solving conflicts. Another subtask of «Evaluate» consists of controlling the design process overall. The group decides which object (requirements or artifact model) can be modified in the next process cycle and in which part modifications can be made («Decision» in Figure 3).

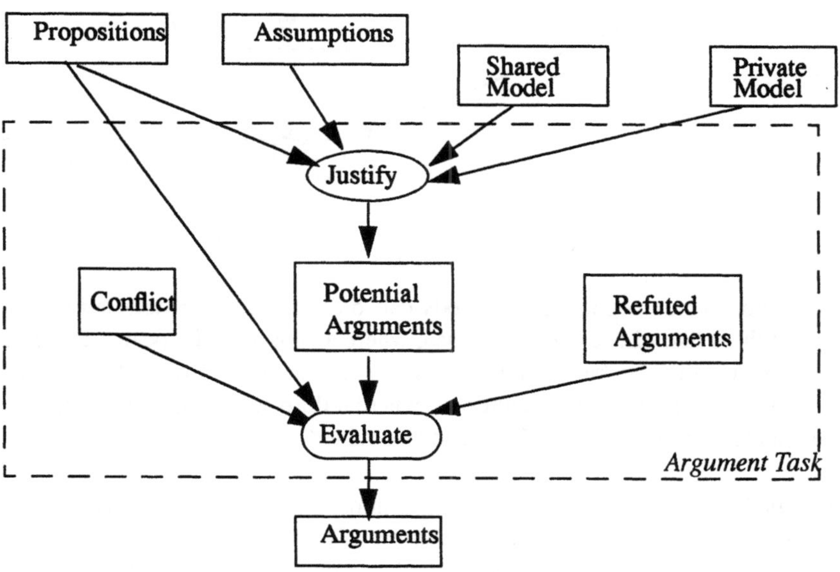

Figure 5. The «Argument» task Model.

Conflict management task consists of first detecting a conflict (its nature) and then solving it. As conflict detection is close to a diagnosis problem, diagnosis methods like those presented in [1] can be used to detect conflicts. Conflict symptoms can be first detected and contributors may be transformed in hypothesis to order to define the elements and the nature of the conflict.

Once detected, to solve a conflict, a solving method is selected and applied. The result is then evaluated to determine if the conflict is solved or not. In the last case, another method can be selected and applied and so on (Figure 6.).

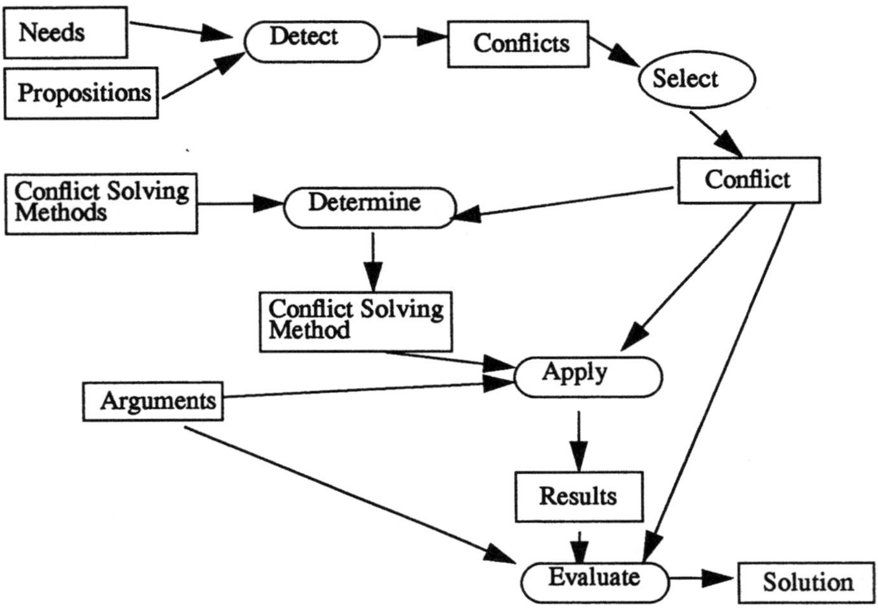

Figure 6. Conflict management task Model.

A large number of negotiation methods can be used to solve detected conflicts. Some of them like «Locate a consensus», where a consensus is located and a solution is then chosen, can be applied to solve mixed-motive conflict [11]. Another like «Introduction of a Third party solution», where a third party imposes a solution [7] is defined to solve resource conflicts and also mixed-motive conflicts.

Other methods are proposed in some conditions. For example, to retrieve conflict solution by applying CBR [14], first a number of past cases must be stored and second, at least one case similar to the current one must be selected. Another method proposes to restructure the problem, using a Goal Graph Search to add goals or to substitute rejected goals [14]. This method needs goal graphs organized as influence trees, to be applied. Defining a «Counter-proposition» or modifying rejected proposition [14] are also recommended.

Given this CE task model, a structure to support this task is determined. We describe its main elements in the next sections.

II. A Structure for Represented and Evaluated Alternative Propositions (SREAP)

In this part, we introduce a definition of a structure to support alternative propositions in CE task. First we define the context of this structure. Then, we describe its basic elements. Finally, we discuss different basic functionalities to use it.

II. 1. Context of the SREAP

Firstly, we have to define what kind of environment we deal with. And, more precisely, where and when SREAP can be used, given the CE task model described above.

Before presenting our structure, let us give an overview of the study in this area. We can note the classification proposed by A. Molina [9] based on technological requirements to develop CE systems. Four types of requirements are needed:

- *Modelling methodologies*: There is a need of modelling methodology, which tackles human behaviour modelling.

- *Computer-aided decision support*: Decision support software tools is necessary to guide designers activity in the whole CE life cycle.

- *Information architecture*: An open and distributed architecture is needed to guide task achievement and communication.

- *Framework for CE environment development*: Environment and tools to build a CE system are needed.

As we have said before, we suppose that the product is developed by a team of designers. All the designers are distributed among a network and the group of designers can evolve for the need of a particular design. So, SREAP can be viewed as a service of an Information architecture.

We can find in the literature examples of such architectures. We present shortly some of those architectures:

- MONET (Meeting On The NETwork) [13] is a CERC (Concurrent Engineering Research Center, West Virginia University) development based on DICE (DARPA Initiative in Concurrent Engineering) architecture. MONET provides a support, and makes transparent the access and the exchange of multimedia data (text, audio, graphics, video and CAD data) and shared computer applications among people over a computer network. Those exchanges of multimedia data can be made in real-time, so MONET is a real-time multimedia conferencing system.

- MIT DICE (Distributed and Integrated environment for Computer-aided Engineering) [12] has been developed at the IESL (Intelligent Engineering Systems Laboratory) at the MIT (Massachusetts Institute of Technology). MIT DICE proposes a communication system among computer and users and a global database implemented as a blackboard.

- PACT (Palo Alto Collaborative Testbed) [6] is a concurrent engineering infrastructure developed at the Lockheed Palo Alto Research Laboratories. PACT is based on a multi-agents system. The agents encapsulate engineering tools and framework and communicate through a shared representation of the design.

Those information architectures tend to respond to the requirement of a good communication between the geographical-dispersed users (human or computer), a shared representation of the design, a design history service, tools and framework application integration services and multimedia user interface [9].

In those information architectures, the object-oriented representation is very helpful to represent the design product and all that is related to it. We do the same for SREAP (i.e., we use the object paradigm for all our descriptions).

As it was notified in [4], we also describe the design of a product in three parts: the product, the requirements and the design process.

In our review of CE systems, we think there is a lack of guidelines for the CE task. Information architectures help to share, store and communicate design information, manage the history of the design process, integrate the design tools and provide an interface for the designers, but they do not help designer to choose the pertinent information to communicate to other designers. Those pertinent information will be evaluated by the designer team.

We will now describe the goal of SREAP. We want to provide support to the cooperative activity of designers. The designers will be helped in their choices between design propositions. Our goal is not to propose an automatic choice between propositions, but to help designers in their choice.

SREAP should be able to support the description of some parts of the product design and the notion explained in our CE's model showed in the part I. Our structure helps to formulate a design proposition and to evaluate it against others.

We will first present the basic principles of SREAP and, then, its use. SREAP will be implemented as a multi-agents system, but we do not describe in this article this implementation. We describe here a structure to support the different activities that those agents will have to help the designers in their cooperative activities.

III. Dealing with Alternative Propositions

As described in the CE model (I), several propositions are made by designers and communicated to the group. The group is composed of specialists of several special fields, and each designer has his specialized domain. Our structure can be viewed as a way to describe and represent the alternative propositions defined in order to evaluate them and to select a solution.

One of the goals of SREAP is to represent all the propositions in a similar structure. Each designer generates his private design and the propositions are extracted from sub-parts of this design and communicated to the other designers.

To illustrate the main concepts used and formalised in SREAP, a short example of designing a room is proposed. There are three designers: an architect, an electrician and a windows provider. Suppose the customer want to make a light room with a window. The requirements associated were the size (between 15 and 20 m^2) of the room and the orientation of the window (not at the north). This stage corresponds to a common design statement.

After this common design statement stage, each designer plans his work to respond to this design problem. Based on their private knowledge the designers communicate their proposition(s). For example, the architect designer proposes a 17m^2 room and a north-east window orientation or a 18.5m2 room and a south window orientation, the electrician proposes electrical fittings for lamps for a room of 15 m^2 and the windows provider proposes a wood casement window for a room of 20 m^2. Each proposition corresponds to an alternative proposition of design for the size of the room.

Alternative propositions must be evaluated against the common design statements. In the evaluation conflicts can appear between alternative propositions. For this evaluation stage, design statements and alternative propositions must be formulated in a standard way, which is what SREAP proposes below.

III. 1. Design Statement and Alternative Proposition

A proposition represents a set of objects proposed by a designer. The proposition is based on the designer's private knowledge (Private Model) and can be evaluated by the other designers. A designer can make several propositions for a design. A proposition is a partial alternative description of a design. In SREAP we call ***alternative proposition*** the representation of a proposition. All alternative propositions corresponding to the same design are linked in SREAP with the associated ***design statement***.

A common decision (within the design team) must be made on what set of objects designers will propose modifications. This set of objects describes the static aspects (represented by objects which describe the product and the requirements) and the dynamic aspects (represented by objects which describe the design process) of the design. We call those objects, ***design statement aspects***.

Here we define what kind of objects are represented in design statement aspects. They can describe the product, the requirements and the design process.

- The description of the design product concerns all the elements which describe and define the product in an intermediate step of design. It is an incremental description, so during the design we have a partial description of the design and by adding product objects we refine and complement the description of the product. For example, the product describes a room which is composed of the window, the electrical fitting, the lamps and so on.

- The description of the requirements concerns all the constraints and elements, that the designer(s) wants to have influence on the product. Requirements are individual or global, they depend on the context of the design description, and they can be hard (indisputable) or express a preference: for example, preference between two orientations of the window in the room.

- The design process deals with the evolution of both descriptions of requirements and product. It characterises the relationship within and between those two descriptions. For example, the design process can provide the requirements which are associated to a particular subset description of design.

Based on those design statement aspects, and his design, the designer communicates an alternative proposition to the designer group. A set of objects composes the alternative proposition. We call this kind of set of objects, an ***alternative proposition description***. The alternative proposition description is represented by a type and the objects which refer to this type. The different types of alternative proposition description are described below.

III. 2. Types of Alternative Proposition Description

An alternative proposition description is represented by the type of elements needed to describe an alternative proposition or to defend it. Each type of alternative proposition description is associated with set of objects. The alternative propositions are described in three general types: aspects, arguments and links.

- ***Aspects:*** Alternative proposition description aspects are represented by objects which describe elements of design in order to refine and to complete the description of the design statement aspects. Aspects adding to the proposition can help to make the choice between alternatives or to define an element of a future design statement. Aspects can be static (i.e. describe the product and the requirements) or dynamic (i.e. describe the design process) as in the design statement aspects. Those aspects can be versions of objects representing design statement aspects or of other objects. We call the version of an object a modification or an addition of properties of an object.

- ***Arguments***: The goal of arguments is to defend the alternative proposition. They are split in two parts: the facts and the functions. In the following, we illustrate arguments with examples retrieved from [14].

  Facts: Arguments are represented by objects. There are two kinds of facts, facts which are parameters of argument functions and facts which describe the qualification of the current proposition (for example the evaluate cost of the proposition). For example, facts can express:

  - ***Goals***: Possible future states of design if the alternative proposition is chosen by the group (for example, used in the generation of threats or promises as arguments).

  - ***Criteria:*** Elements used for qualification of the alternative proposition (for example, used in the appeal of a criteria corresponding to a specific theme or in the prediction of the future by extrapolating currents facts).

  - ***Examples / Counter examples***: References to past cases, contradiction points of another alternative proposition (for example, used in definition of counter examples as arguments or in the appeal of exceptions retrieved from minor standard justification).

  - ***Believes or Behaviours***: General uses (or habits) of designer group (for example, used in prevailing practice justification).

  Functions: Functions describe (alternative) methods to calculate the interest of the proposition. For examples, methods can be:

  - ***Case Based Reasoning***: Methods to express arguments based on past cases.

- ***Preference Analysis***: Methods to express an ordering between alternative propositions (for example, used in prevailing practice justification or in the appeal of self-interest argumentation).

- ***Rules/Heuristics***: Methods to express some general principles (for example, used in universal principles justification) or exceptions (for example, used in the appeal of exceptions retrieved from minor standard for the argumentation).

- ***Links:*** Always, an object (representing aspects or arguments) is linked with others. For example, a structural link (for example the link «A is sub-part of B», «C is composed of D,E,F»), a functional link (for example the link «A is equal to B», «C is greater than D»), a logical link (for example the link «A implies B»), or a composition of links. Thus, we can describe different types of links between objects and for each link, we can have the inverse link. So, we view link as a composition of objects defined later (which depend on another design decision) or already defined somewhere else.

III. 3. A structural View of the SREAP

SREAP can be viewed in structural way like a three levels tree: design statements, alternative propositions and alternative proposition descriptions. In the following, we call it SREAP tree, in which:

- design statement is the root of the tree,

- alternative propositions are sons of the design statement,

- alternative proposition descriptions are the leaves.

We associate a list of objects to a design statement. A design statement is labelled by a reference name or by a name chosen by the designers who have defined the design statement aspects.

An alternative proposition is a set of alternative proposition descriptions. An alternative proposition is labelled by a name selected by the designer who have made this proposition.

The alternative proposition description is composed of a type and a set of objects which defines and characterizes the alternative proposition. An alternative proposition description is labelled by the name of its type.

After defining the three different levels of the tree, we define below the relationship between sons in a tree. There are two kind of relations: alternative and associative. The alternative relation is between alternative proposition nodes and the associative is between alternative proposition descriptions leaves. The alternatives relations can be viewed as an «or» relation between alternative propositions and the associative relations as an «and» relation between alternative proposition descriptions.

An example of an SREAP tree is proposed in Figure 7: the design statement «Luminosity» defines the luminosity of a room, the object in design statement aspect is «Light». Two alternative propositions are associated to this design statement, labelled by «Good» and «Half-light». In the Good alternative proposition, we refer to a design statement named «Artificial_Luminosity» (Design Statement = Artificial_Luminosity), in this case we argue that the artificial light source provides a good luminosity (Fact:= Light.quality = «Good»). In the Half-light alternative proposition, the window must have a size between 1 and 2 m2 and must be defined in alternative «Half-light» of design statement «Window_Luminosity». In this alternative proposition the design statement object Light is refined.

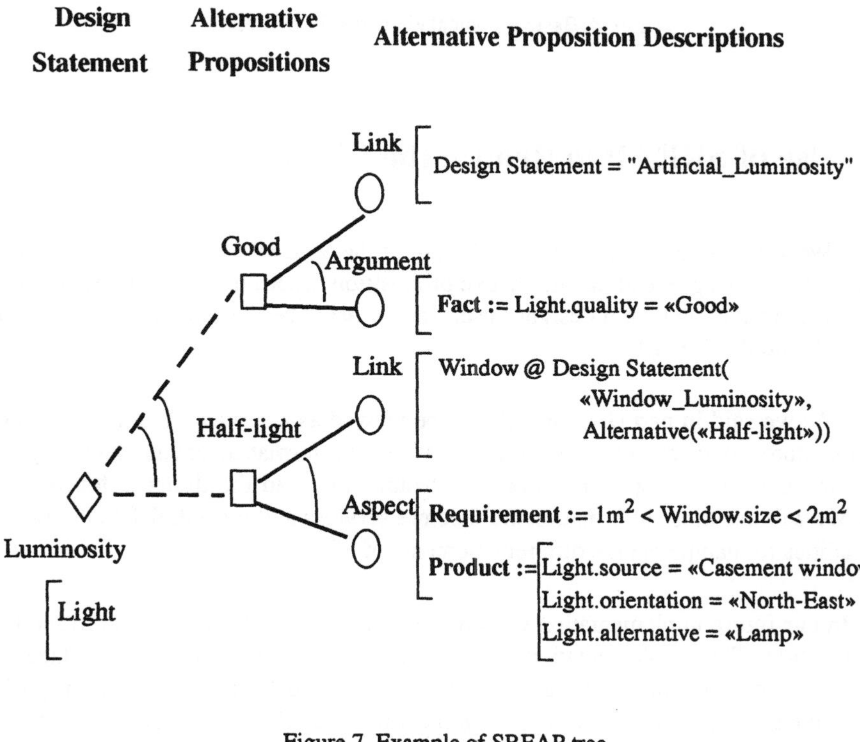

Figure 7. Example of SREAP tree.

III. 4. A representation view of the SREAP

For a more ergonomic use of SREAP we have associated a graphical representation of each elements of the tree. So, the different elements of SREAP are represented in Figure 8.

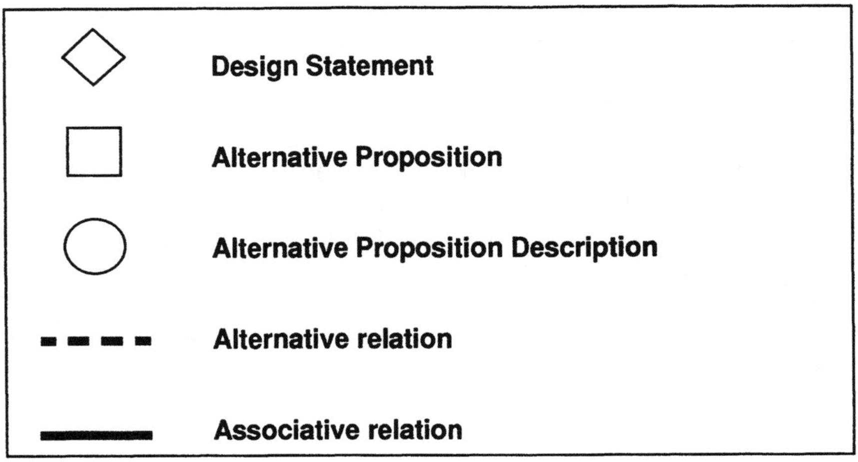

Figure 8. Basic graphical elements for SREAP.

IV. Use of SREAP to Realize the CE Task

We will define the meaning and the use of the SREAP elements (design statement, alternative proposition, alternative proposition description and alternative and associative relations) for a designer use. So, we describe the relationship between SREAP and the CE task.

As we said in part (II), SREAP can be viewed as a service of an information architecture, so we dispose of the other services of information architecture. In particular, designers using information architecture can read the design information. So, we will use those information architecture services and we will define the functionalities to manipulate the different elements of SREAP tree.

In our future implementation we will provide an aid to formulate and to send a proposition. This implementation will propose some services as user interface to write and to manipulate the elements of his sub-part of design, read the propositions, communication tools to support proposition and information sending between designers and help to manage conflicts between propositions. In this article, we focus on SREAP functionalities that are needed to describe CE task proposed in part (I). SREAP functionalities must help designers to manage, share, store and communicate alternative propositions and design statements. For this purpose, we define some basic functionalities Read, Write and Send. In the sections below, we show the use of those functions in a CE task. We only specify the general definition of those functions. We do not present in this paper how SREAP can support tasks realization and conflict management.

- **Read**: The elements of a SREAP tree must be accessible to a designer. The designer can: ask elements of private or shared SREAP tree, update elements of his private SREAP tree from the CE environment or his private resource, or read in a shared SREAP tree elements for his private SREAP tree.

- ***Write:*** A designer must be able to define elements of a SREAP tree. The designer can insert / delete elements in his private SREAP tree.

- ***Send:*** The elements of a SREAP tree must be exchanged between designers. The designer can: tell to other designers, or insert / delete in a shared SREAP tree elements of his private SREAP tree. Or elements of shared SREAP tree can be communicated to a designer.

In the following, we present the use of the SREAP in communication between designers in a CE task. For each element of a CE task description we define what are the elements of the SREAP used. To sum up the use of SREAP functionalities we provide figures for each CE task.

IV. 1. Design Task

In a first step we have to help each designer to form his propositions. For each designer we provide SREAP tree to help him to formalize his proposition, and so on for all designers. Each sub-part of design corresponds to a private and partial part of design. The proposition will correspond to the description of an alternative proposition in a shared SREAP tree.

- ***Artifact, Private model*** and ***Shared model***: In SREAP, we are interested in the exchange of information between the designers, so all the parts which correspond to private or synthesis representation are not represented in SREAP. In particular, *Artifact*, *Private model* and *Shared model* are not represented in an extensive way in a SREAP tree, but elements of *Artifact*, *Private model* and *Shared model* can be used to describe elements of a proposition in a SREAP tree. In the rest of the document, we assimilate elements retrieved from the *Artifact*, the *Private model* and the *Shared model* which are represented in a SREAP tree as, respectively, *Artifact*, *Private model* and *Shared model*.

- ***Requirements***: Description of requirements, which are used in a proposition, are represented by design statement aspect requirements or by alternative proposition description aspects of shared SREAP tree.

- ***Explicit Requirements***: They are represented by alternative proposition description aspects of the actual designer's private SREAP tree, if they are used in a proposition.

- ***Violated requirements***: They are represented by design statement aspect requirements or in alternative proposition description aspects of one, or more, shared SREAP trees, if they are used in a proposition.

- ***Sub-part of Design/Assumptions***: Elements of *Sub-part of Design/Assumptions* are represented in an alternative proposition of an actual private SREAP tree, if they are used in a proposition.

- ***Propositions/Assumptions***: They are represented by alternative propositions of the actual private SREAP tree. *Proposition/Assumptions* are chosen to be sent to the other designer and they are represented in a shared SREAP tree.

- ***Specify***: The goal of the sub-part of design is to provide a more precise and explicit list of requirements from the set of requirements provided to the designers. Designers can be helped by using the shared SREAP tree to formulate the global requirements of the product to design in the design statement. Reading the definition of the global constraints and the common description of the current design helps to express the constraints in order to be more specific and compatible with the global and private (described in the private model) constraints, and description of the past and current description of the product (described respectively in the artefact and in Shared model). SREAP will help to realize this task by providing services of Read of the *requirements*, Read of the *Shared model and* Write *explicit requirements* (cf. Figure 9).

- ***Generate***: To satisfy the explicit requirements, the designer builds a *sub-part of design/assumptions*. The designer, in his private work, could have several partial descriptions of his final sub-part of design. Or, he can have several sub-parts of design to propose to the other designers. So he can formulate a set of sub-parts of design as alternative propositions of his actual private SREAP tree. SREAP provides a common requirement (satisfied or violated) representation. So, the designer is helped to generate a *sub-part of Design* using some services provided by SREAP like: Read of the violated *requirements*, Read of the *Shared model* Read *explicit requirements* and Write the *sub-part of Design/Assumptions* (cf. Figure 9).

- ***Evaluate***: A designer has to evaluate if a *Sub-part of Design/Assumption* satisfies the *explicit requirements*. The unique representation of *Sub-part of Design* and *requirements* (in alternative proposition description aspects and design statement aspects) guides the designer to evaluate if the *Sub-part of Design* satisfy the *requirements* and to detected violated ones. SREAP will help to realize this task by providing services of Read of the *Shared model*, Read *Sub-part of*

Design, Read *explicit requirements* and Write the *violated requirements* (cf. Figure 9).

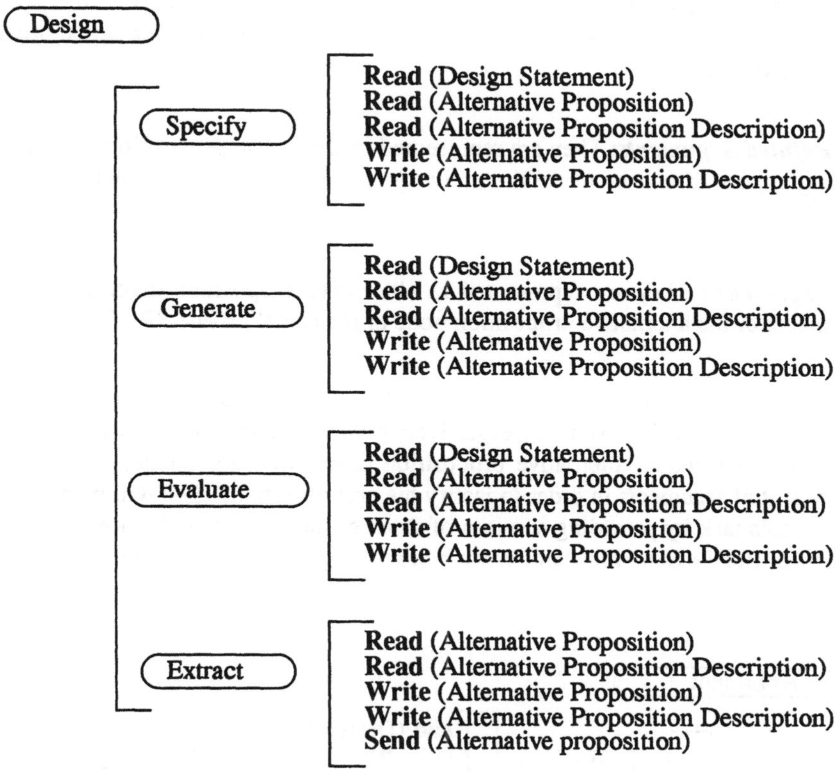

Figure 9. Design task.

- ***Extract***: When a designer has finished to formulate his *Sub-parts of Design*, he has to choose one of his alternative propositions (corresponding to *Sub-parts of Design*) of his actual private SREAP tree and to send a proposition to the group. From a *Sub-part of design* is extracted a proposition and is stored as an alternative proposition in shared SREAP tree. SREAP will help to realize this task by proposing elements which are potentially represented in a proposition as an alternative proposition of a shared SREAP tree. We provide services as: Read *Sub-part of Design*, Write *proposition/Assumptions* and Send of the *Proposition/Assumptions* (cf. Figure 9).

IV. 2. Argumentation Task

Communication in the argumentation task (part I) can be handled as follows.

- ***Conflict***: The source of conflicts is represented within the alternative propositions of shared SREAP tree.

- ***Potential Arguments***: They are represented in alternative proposition description arguments of one or more alternative proposition nodes of the actual private SREAP tree.

- ***Refuted Arguments***: They are represented in alternative proposition description arguments of one or more alternative proposition nodes of the actual private SREAP tree.

- ***Arguments***: They are represented in alternative proposition description arguments of the alternative proposition node of the actual private SREAP tree.

- ***Justify***: Based on his *Assumptions* and *Proposition,* represented as alternative propositions in SREAP tree (respectively private and shared), a designer defines, and represents as alternative proposition description arguments, a number of potential arguments in order to justify his proposition. SREAP will help to realize this task by providing services of Read of the *Shared model*, Read *propositions*, Read *assumptions* and Write *potential arguments* (cf. Figure 10).

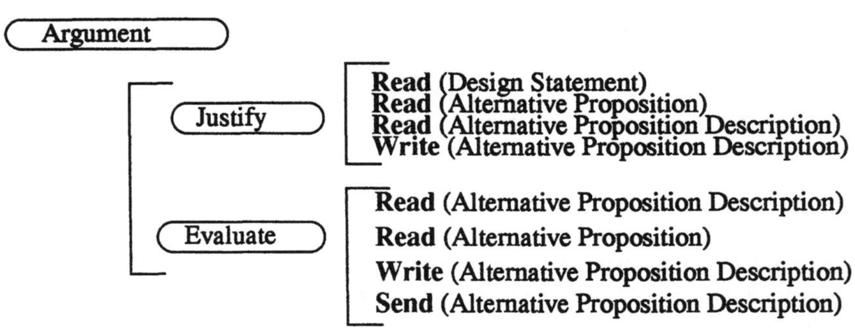

Figure 10. Argument task.

- ***Evaluate***: SREAP will help to realize this task by proposing *Potential or Refuted Arguments* which is selected or modified (represented as an alternative proposition description arguments of a shared SREAP tree) considering conflict sources described in alternative propositions. We provide services as: Read of the *conflict*, Read of the *propositions*, Read of *the potential arguments*, Read of the *Refuted Arguments*, Write the *Arguments* and Send the *arguments* (cf. Figure 10).

IV. 3. Evaluate Task

As described in part I, the evaluate task is decomposed in conflict management and in design process control.

IV. 3. 1. Conflict Management Task

- ***Needs***: They are design statement aspect requirements or alternative proposition description aspects requirements of shared SREAP tree.

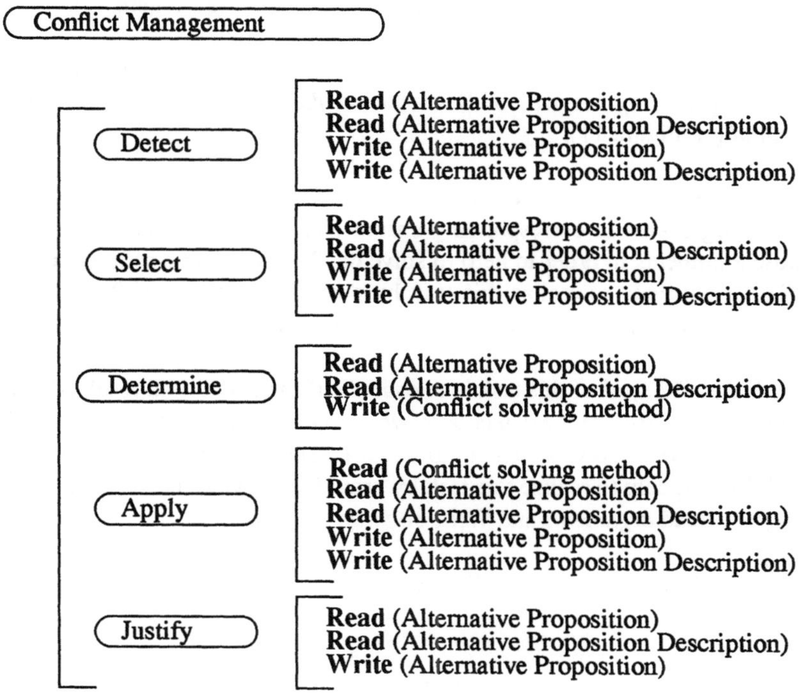

Figure 11. Conflict management task.

- ***Conflict Solving Methods***: They are associated to a SREAP tree. They are not represented in the tree, because they are methods between propositions, and we represent only elements of propositions. In our future work, we plan to use those methods with SREAP and the way to represent those methods is not already defined.

- ***Results***: They are represented by alternative proposition nodes of a shared SREAP tree.

- ***Solution***: They are represented by alternative proposition nodes of a shared SREAP tree.

- ***Detect***: The goal of this task is to detect what element of which proposition violates a need represented as a shared requirement. SREAP represents both requirements and proposition in shared trees. So, the detection of conflict sources is guided; a conflict source is an alternative proposition description which does not satisfy the needs. SREAP will help to realize this task by providing services of Read the *needs*, Read *propositions* and Write the *conflicts* (cf. Figure 11).

- ***Select***: It is a choice of a *conflict*. SREAP will help to realize this task by providing services of Read of the *conflicts* and Write *conflict* (cf. Figure 11).

- ***Determine***: It is a choice of a *conflict solving method*. SREAP will help in this task by providing services of Read the *conflict* and Write a *Conflict solving method* (cf. Figure 11).

- ***Apply***: It is the application of the *conflict solving method*. The possibility to represent arguments in corresponding shared alternative proposition helps to use those arguments in order to help to solve conflict and define solution as results. SREAP will help to realize this task by providing services of Read a *Conflict solving method*, Read *conflict*, Read *arguments* and Write *results* (cf. Figure 11).

- ***Evaluate***: The goal of this task is to evaluate if the results had resolved the conflict selected. So, like the detect task, we search what element of results violate a need. SREAP represents the both requirements and results in shared trees. SREAP will help to realize this task by providing services of Read *results*, Read *conflict*, Read *arguments* and Write *solution* (cf. Figure 11).

IV. 3. 2. Control of Design Process Task

Figure 12. Control of design process task.

- ***Decision***: The goal of this task is to express which objects (which describe product and requirements) will determine the design statement aspects in the next process cycle. Future design statement aspects will complete and refine the current solution by add of new alternative propositions. SREAP will help in this task by providing services of Read of the *Solution*, Read of the *Requirements*, Read of the *Shared model and* Write *Decision* (cf. Figure 12).

We have shown how to represent and share an alternative proposition with respect to the CE model described in part I. All individual actions supported by SREAP, are not static; the designer task can not be reduced to the use of a simple representation of an alternative proposition node in a SREAP tree. In order to help a designer, we have proposed a set of functions corresponding to actions that designers do in respect of our model propose in part I. SREAP propose functionality for load, store and send pertinent information. Those pertinent information are structured in a single feature in order to make conflict detection and solving easier.

Conclusion

Concurrent Engineering is a complex task in which individual competencies are confronted. The objective of our work is to provide guides to help the modelling of this task and to support its realization.

We proposed a generic model in which individual competencies and common ones are handled. Our model is general and can be used in different types of pragmatic design cases. The decomposition of a CE task in three level reflects the real-life realization of this task and allows to control methods used to perform it and to manage conflicts detected between designers. Our decomposition makes explicit the operation nature in CE, so a best management of the elements which implements those operations can be reached.

We also define a structure to support CE task representation. Knowledge management and representation are held in this structure.

This presentation of our work corresponds to the first step of the definition of SREAP. In this article we describe a way to represent pertinent information in a part of the CE task. In this representation we have provided a structure to help designers to formulate a proposition in order to be understood by others (we suppose that the designers use the same language and the same ontology, yet we do not manage the problem of misunderstanding due to the different technical vocabularies).

In order to help a collective choice between propositions, SREAP must support collective exchange and represent pertinent information. For the collective actions and representation, in respect of the CE task described in part I, we use the same structure and functionality: we represent only the pertinent information (sent proposition) and use the same set of SREAP functions. We focused on the management of disagreement between propositions. SREAP could help to define sources of conflict (as a part of a proposition) in the process of conflict management. Here we do not express the conflicts management methods nor the conflict types identification.

Our structure, SREAP, proposes help by providing guidelines which orient designers in their CE task through a structured communication and by providing a common representation of the deferents manipulated CE proposition elements. We feel there is a lack of such guides in information architecture.

Some parts of a CE task are not represented in SREAP because they are not pertinent to the exchange of information between designers. The past design descriptions are not represented in SREAP. To represent and read private and past information, we can use technologies used in information architecture.

The need to represent arguments [14] generated with propositions concerning the product generated and the requirements manipulated [4], pushed us to group these elements in a unique structure.

In order to guide CE and conflict management task modelling, we are going to define a generic conflict management methods library and to study conflict types as index of this library. We plan to implement this SREAP in a multi-agents system. We will study the possibility of detection of the different nature of potential conflicts in terms of SREAP and the application of a conflict solving method. A definition of a scenario of CE design and a validation by users will show the interests and problems of SREAP.

Acknowledgements

We deeply thank the French «Ministère de l'Enseignement Supérieur et de la Recherche» (Genie Project) that funded this research and the ModelAge Working Group (contract EP:8319) for their partial funding. We thank also Dr. R. Dieng and Dr. O. Corby for their remarks.

References

1. Benjamins R 1993, Problem Solving Methods of Diagnosis, Thesis Univeriteit van Amsterdam, Amsterdam.
2. Bond A H 1990, A Computational Model for organizations of cooperating intelligent agents, *Proceedings of the Conference on Office Information Systems*, Cambridge.

3. Boyera S. , Corby O. 1995 *Etat de l'art sur la multi-expertise, Projet Génie Thème 3,* Rapport de contrat GENIE, Thème 3, Sophia Antipolis.

4. Brazier F M T, Van Langen P H G , Treur J 1995 *Modelling conflict management in design: An explicit approach,* Artificial Intelligence for Engineering Design, Analysis and Manufacturing, Vol.9, N.4, Cambridge University Press, USA, pp.353-366.

5. Breuker J, Van de Velde W 1994 *CommonKADS Library for expertise modelling Reusable problem solving components, Frontiers in Artificial Intelligence and Applications,* J. Breuker and W. Van de Velde (EDS), Amsterdam: IOS.Press 1994.

6. Cutkosky M R, Engelmore R S, Fikes R E et al. 1993, «PACT: An Experiment in Integrating Concurrent Engineering Systems», *IEEE Computer*, pp. 28-37.

7. Easterbrook S M, Beck E E, Goodlet J S, Plowman ., Sharples M, Wood C C 1993 A Survey of Empirical Studies of Conflict, *CSCW: Cooperation or Conflict?* S. Easterbrook (ed), Springer-Verlag.

8. Klein M 1995 Conflict management as part of an integrated exception handling approach, Artificial Intelligence for Engineering Design, Analysis and Manufacturing, Vol.9, p.259-267, Cambridge University Press USA.

9. Molina A, Al-Ashaab A, Ellis T I A, Young R I M, Bell R 1995 A Review of Computer-Aided Simultaneous Engineering Systems, *In Research in Engineering Design*, vol. 7, number 1.

10. Newell A 1982 The Knowledge level, *Artificial Intelligence Journal,* 19 (2).

11. Ramesh B, Sengupta K 1994 Managing Cognitive and Mixed-motive Conflicts in Concurrent Engineering, *Concurrent Engineering: Research and Applications* Vol.2, N.3, pp.223-236.

12. Sriram D, Logcher R, Wong A, Ahnmed S 1990 An Object-Oriented Framework for Collaborative Engineering Design, in D. Sriram, R. Logcher and S. Fukuda (eds.) *Computer-Aided Cooperative Product Development*, Springer pp 51-92.

13. Srinivas K, Reddy R, Babadi A, Kamana S, Kumar V, Dai Z 1992 CERC-TR-RN-91-009 MONET: a Multi-media System for Conferencing and Application Sharing in Distributed Systems, *CERC Technical Report Series*, CERC-TR-RN-91-009, Morgantown, WV: CERC, West Virginia University.

14. Sycara K P 1990 Persuasive Argumentation in Negotiation, *Theory and Decision,* 28: 203-242, Kluwer Academic Publishers, Netherlands.

15. Sycara K P 1991 Cooperative Negotiation in Concurrent Engineering Design, *Computer aided cooperative product development. Proceedings of MIT-JSME workshop*. D. Sriram, R. Logcher, S. Fukuda (Eds), Cambridge, MA.

16. Wielinga B , Van de Velde W, Schreiber G, Akkermans H 1992 *The CommonKADS Framework for Knowledge Modelling,* Report of ESPRIT KADS-II project p.5248: KADS-II/1.1/pp/UvA/35/1.0.

Part IV
Design Knowledge & Information

Design Information Issues In New Product Development

O.P. Boston[1], A.W. Court[2], S.J. Culley[1], C.A. McMahon[3]

ABSTRACT

The need for engineering designers to use relevant and up-to-date information is a crucial issue for companies concerned with the design and development of new products and systems. The pressures to innovate and provide new products that are of increased quality, reliability and performance has never been higher.

The purpose of this paper is to investigate the key issues in the domain of the provision of information for engineering designers. This will be achieved by analysing the characteristics that are special to information in the engineering domain, including: the importance of obtaining information by informal means; the key role of suppliers; and the extent to which engineering designers rely upon their own personal knowledge.

1. INTRODUCTION

Within the design and development of new products that are of increased quality, reliability and performance, the need for engineering designers to use relevant and up-to-date information has become a major influence and priority for many individuals and companies. Every day engineering designers use information from a variety of sources to undertake a wide range of design tasks. Without access to accurate and up-to-date information they may make mistakes or misjudgements on aspects of the products design, as described by Benyon [1]. For example designers may not be aware of the latest manufacturing capability of the company, supplier or subcontractor, the design may not be suitable for final manufacture without revision being necessary to take account of discrepancies for specific machining processes [2]. It has also been shown that engineering designers spend as much as 30% of their time searching for and accessing engineering design information [3] [4] [5]. To try and reduce this 'non-productive' time engineering designers tend to use the information that they already possess [6]. This may result in designs being generated without the benefit of information that <u>does</u> exist within the enterprise, but is too time consuming to find, or exists within the domain of the supplier and is even more time consuming to track down. This ultimately must lead to a reduction in

[1] School of Mechanical Engineering, University of Bath, Claverton Down, Bath, BA2 7AY. Tel: +44(0) 1225-826129 Fax: +44(0) 1225-826928

[2] Department of Mechanical Engineering, Imperial College of Science, Technology and Medicine, Exhibition Road, London, SW8 2BA. Tel: +44(0) 171-594-7050 Fax: +44(0) 171-8238845

[3] Department of Mechanical Engineering, University of Bristol, Queens Building, University Walk, Bristol. BS8 1TR. Tel: +44(0) 117-9-288100 Fax: +44(0) 117-9-288912

productivity and worse still crucial design decisions will be based on incomplete data and assumptions and they are therefore likely to be sub-optimal.
Benyon [1] suggests that *"Information is such a familiar concept that we rarely think about it"*. This over-familiarity may go some way to explaining the low priority given to organising and managing information for engineering designers. There are two views of information to be considered, one is the classical information theory approach of Shannon and Weaver [7], that it is "the relative frequencies of signals emanating from a source". The second, of more relevance, is that it can be considered as "a measure of its usefulness to a specific task being performed". The work of Pahl and Beitz [8], Rouse [9] and Rzevski [10] have been directed towards establishing measures for information usefulness, which may be categorised into the following areas of interest:

- Medium of presentation.
- Format of presentation.
- Location of delivery.
- Control of delivery.
- Up-to-dateness.
- Timeliness.
- Accuracy.
- Relevance.
- Cost.

These categories will be used as the basis for the first section of this paper. The paper then focuses in particular on the importance of obtaining information by informal means, the key role of suppliers and the extent to which engineering designers rely on their own personal knowledge and the implications that these findings have for the future provision of information in the product development process.

2. THE PROVISION OF DESIGN INFORMATION IN NEW PRODUCT DEVELOPMENT

The two elements in the middle of the above list 'Control of delivery' and 'Up-to-dateness' have been combined under the title of 'Management of delivery'. This is considered a sensible revision to reflect the changes that have taken place in recent years.

2.1 Medium of Presentation

The success of the innovation process in new product development depends on functional specialist team members bringing to bear knowledge and resources needed to achieve the team goal [11]. Information must be processed between them to provide the co-ordination needed for high performance. Today, this sharing of information can be achieved via a wide variety of communication media, such as: face-to-face; video conferencing; e-mail; memo; formal document; etc.. The selection of which, is frequently within the control of the engineer, but once chosen it dictates the richness of information that can be processed, as shown below in Figure 1. Yet the authors have noted that the engineer is unlikely to have received any formal guidance on selection of the media required to process information of the appropriate richness to reduce uncertainty and clarify ambiguity.

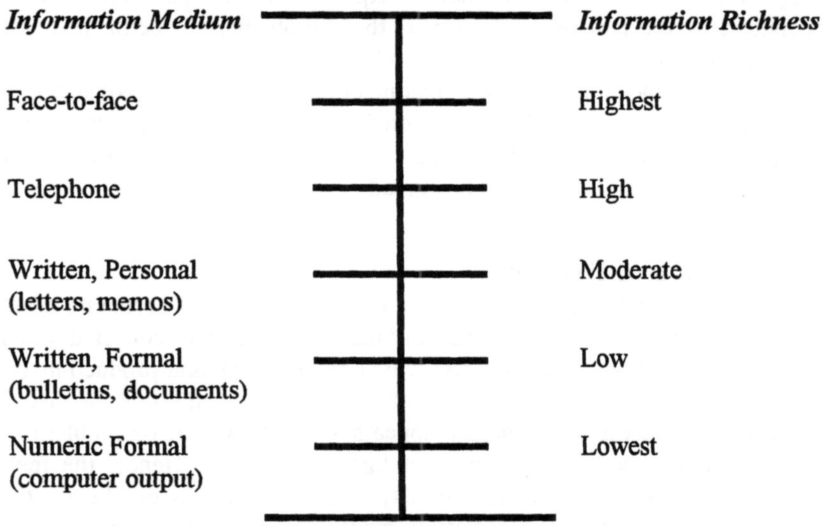

Figure 1: Communication Media and Information Richness [12].

Interpersonal channels of communication are important when perceived uncertainty is high [13], as high rich media are better than low rich media at eliminating ambiguity [12]. Hence, the use of such communication channels is particularly important in the preliminary stages of design, and in this context Daft and Lengel [12] have reported that face-to-face communication within an organisation is often easier to gain access to than other methods. However, the findings of Liker and Hancock [14] throw a different light on this, as they found in their study of a large automotive company that attempted contacts were successful only 32 % of the time. Thus, it may be concluded that extensive face-to-face meetings are seldom practical [15], for simple phenomena, they may be inefficient, and facial expression or tone of voice may distract from spoken words [12]. Therefore, once differences have been resolved and agreement has been reached, less rich forms of communication media such as faxes or e-mail should be used.

2.2 Format of Presentation

Information can be presented in a multitude of formats and languages, and this greatly affects its future applicability and usefulness. For example, Turner [16] reported that well produced diagrams, such as those in some supplier catalogues, give clues via high impact visual representation and are particularly useful for 'sparking' ideas. The format of the information within catalogues is however beyond the direct control of the engineer, but even when it is difficulties arise because they frequently transfer information in the format that is easiest for the

originator. These problems also extend to electronic data, as the existence of many types of software and various methods for data representation implies that data provided by one source is likely to require some manipulation to the format understood by the end user. However, these problems are being addressed by initiatives such as CALS [17] and standards such as STEP [18].
It may be concluded that current media do not facilitate rapid access across organisational boundaries [19], and therefore, prior to product development, an appropriate format for the presentation of information to be shared between a customer and its suppliers should be agreed.

2.3 Location of Delivery

A companies knowledge base is widely considered to be a corporate asset [20] [21]. It is comprised of the information/data that is created, collected and stored on various media types, the processes which guide design and production, and the people within the organisation [22]. It is especially important in mechanical engineering as the essence of design appears to rely very little on problem-solving techniques but rather on a richness of knowledge [23]. Hence, the means of accessing it is of prime importance [24], and this depends on the relationship between the location of the information and the location of the requester [25], and in particular the nature of this relationship during product development. This has been the topic of a great deal of research [26] [27], for example, the authors in a survey of over 200 practising engineering designers [28], reported that when starting a new task the following information sources were rated as Very Important :

60 % Colleagues.
34 % Personal Contacts.
84 % Personal Experience.
17 % Representatives.
12 % Consultancy.

These values are typical of those in the many similar studies that have been undertaken, where generally during product development it has been noted that engineers rely heavily upon their own knowledge, a network of colleagues and a personal collection of supplier information [29] [30]. This information is usually located close to the engineers' elbow [16] [31], and hence to be of value all information should be accessible from this point.

2.4 Management of Delivery

Engineering companies today are inundated with information from a variety of different sources, such as: data created on CAD or word processing systems; paper documents created in-house; and paper or film coming in from suppliers. Here a potentially difficult situation arises for the engineer, as with too much data critical items can easily be overlooked, and with too little the engineer may have to from a conjecture. Also implied is the reference to the designers and their selectivity in receiving and structuring information, which is dependant on their personality [32]. However, this selectivity may have been performed already, especially in large organisations, as the engineer is likely to receive filtered, delayed and edited market information through interface organisations [33]. Unfiltered information flows are however critical to the success of any organisation [34]. Further, some information dates quickly, in which case an electronic format would be advantageous for easy up-dating, but other information dates slowly, and therefore the danger of having only an out-dated hard copy is reduced [35]. Hence, the control and delivery of information can be a nightmare if it is not managed effectively [22].

To address some of these problems, document image management systems have been developed which are capable of storing scanned images of both hard data and soft electronic data. Their implementation within the field of engineering design is gradually increasing, but to be useful, engineering data must remain live such that it actively reflects the current situation, and one of the problems with these systems is updating them. If the amount and quality of information within them is below some threshold, then the engineering designers will gather a significant amount from other sources and the system will be used less frequently [36]. Further, insisting that they use them may be counter productive [37], and there is no guarantee that they will not rely on their own personal collection of printouts that were copied from the system two years previously.

One of the successes which has emanated from the authors' work was the provision of standard component information by what is referred to as the "Electronic Catalogue" [38] [39]. This enables manufacturers to distribute their technical information on computer disc, CD ROM, or even provide access to the most up-to-date versions over the World Wide Web (WWW). In more sophisticated embodiments of this technology the electronic catalogues will also include selection algorithms that automatically perform appropriate calculations and search functions, helping to minimise time wastage whilst ensuring an optimal search.

2.5 Timeliness

There have been numerous studies into the percentage of time that engineers spend on their various activities. Liker and Hancock [14] reported that lack of time was the most important barrier to engineering effectiveness, and found in a study of a large automobile company that engineers reported spending on average only about 8 % of their time actually designing. The results presented by Will [40] also agree with this, reporting that the average engineer spends approximately 10 % of his or her time designing.

These findings may be largely owing to the fact that engineers waste large amounts of time tracking down information [25]. Rzevski *et al.* [10] have suggested that often up to 70 % of their time is taken up with this activity and Yeaple [33] has estimated that these delays may be responsible for increasing the average product development time by up to 48%. Schierbeek [41] noted that availability of means for communication is quite frequently a problem, and this alone could have a disastrous affect on delivery times. Moreover, when designers need access to information they normally need it urgently [42], and if delivery times are unsatisfactory, decisions are made without the benefit of good information [15] [24]. Hence, as Schwarzwalder [29] states, there are tremendous advantages for those who understand the sources of information and can efficiently find what they need.

Within the confines of the company, the database provides an efficient, easily comprehensible framework for the organisation of design information [36]. Typically, compared to manual systems, they can reduce the time taken to access and return a document from two days to two hours [22]. However, improperly designed systems may impede information flow, frustrate communication between people, and they can lead to the duplication of information and the compounding of human errors [21].

2.6 Accuracy

The accuracy of data may concern the physical, experimental or other basis of data, but it may also concern the accuracy of the transcription of values utilised by the engineer. A simple error in a figure or decimal place in a table could result in the catastrophic failure of a product. These errors are not out of the question, as Reynard [43] has reported that books, manuals, journals and the data and information from materials producers and stockists are of varied quality and usefulness, and MacCallum [44] calculated that "the best of the large books of tables of integrals have about 7 % of the formula wrong - the worst may have an error rate of 25 %". Pitts [31] has also noted, that standard catalogues often represent products with an aggressive marketing policy and do not always represent the most appropriate item for the job, both in terms of performance and also cost. These findings are alarming since many engineers rely on design analyses presented as part of suppliers' literature [31] and most will use such information with false confidence.

Not only do concerns exist about the credibility of information provided to engineers, but protocol studies performed by Stauffer *et al.* [45] have shown that often their own decisions are based on rules-of-thumb, personal preferences and even conjectures, which are formed when there is not enough information to know things with certainty, but enough to make an informed guess [46].

Hence, as Almli [47] states, there is a need for tools and methods to *evaluate* the quality of the information used in product development. This being particularly important for any collaborative process as one of the key issues for co-ordination is consistency and integrity of information [48].

2.7 Relevance

To *resolve* uncertainty in the industrial context relevant information is needed, but this uncertainty can also be *reduced* by providing information about information. This means providing leads to locating the information of primary interest required to *resolve* the uncertainty [49]. In this context, purely for the purposes of awareness, the benefits of subscribing to a commercial database now become clear. In 1988, March and Trott [50] reported that over 3000 were in existence, covering management, business, science and technology in a very wide range of specialities. A particular example of such a database is discussed by Reynard [43] as containing data for over 11 000 materials from about 1000 suppliers. The databases serve as directories of services and products, some providing full text printouts of supplier catalogues.

Leads to locating information are also enabled by literature scanning, the single most important reason for doing so was found by Shuchman [26] to be 'keep current in my field', with 70 % of the engineers surveyed rating this as the 'most important' reason. Organisations are also connected to outside sources of technology by the 'gatekeeper' who keeps up with new technical developments by reading more technically sophisticated literature and by communicating with technical experts. Further, as already discussed, because of his proven competence he is frequently consulted on technical matters and as a result, Nochur and Allen [51] have described the gatekeeper as a very effective channel for transferring relavent technical information into an organisation from external sources. However, the gatekeeper should not be solely relied upon as an external link, as Sheen [52] has reported that although they may draw the attention of others to information, ultimately digestion is a solitary process and dissemination is limited.

2.8 Cost

The costs associated with say: obtaining; waiting for; managing; misusing; ignoring or translating information may be practically impossible to calculate in isolation. However, it has been noted that engineers waste considerable amounts of time waiting for information, which results in costly delays in market introduction. For example, Clark [53] has conservatively estimated that in the case of a car that sells for $10,000, each day of delay costs the organisation over $1 million in lost profits. This compares to the estimate of Blotwijk [54] that within Europe paperwork delays owing to manual bureaucracy amounted to 7 % of the value of the goods concerned. It is known that during the early stages of new product development over 80 % of the total project costs can be built in [55], hence in this context, the importance of up-to-date, accurate relevant and timely information becomes clear.

Throughout this paper the computer has been highlighted as the target medium for information management, and Sanford [56] has estimated that if effective it has the potential to generate savings of between 2 and 5 % of total project costs. However, the uptake has been slow, which, considering the potential benefits highlighted, begs the question why ?. Well firstly, assuming that the organisation can afford to purchase or develop an appropriate system, there is the major task of sorting out the existing information and transferring it to the system. Secondly, there is the task of maintaining it which involves considerable changes in working habits of employees. Thirdly, computers are not without their problems, for instance, Bögler [57] has reported that six out of ten big British organisations have suffered a major failure with their computer systems in the past two years.

Finally, people are very concerned about security issues, and among those companies who had suffered problems, 35 % had experienced a breach of security by internal staff or a third party. Security issues are a major concern within industry, and although the employment contract inherently presumes a certain measure of loyalty towards the employer [58], it is likely that security issues alone have had a drastic effect on the uptake of computer based information management systems.

3. THE ROLE OF THE SUPPLIER

The previous section has highlighted some of the issues relating to the provision of engineering design information during the product development process. This section however, deals with issues relating to: the provision of supplier information; the role of suppliers in new product development; and the authors' current work in relation to these areas.

3.1 Suppliers as an Information Source

Allen [59] reported that suppliers have been the most used information source for engineering decisions, and Bond and Ricci [60] describe the use of major supplier part information in the preliminary stages of design. A study by Radcliffe and Lee [61] highlighted the frequent use of supplier catalogues and Schwarzwalder [29] reported that they were an indispensable tool for the working engineer. Stauffer *et al.* [45] noted that they were used when designers possessed little domain knowledge and wanted to find what was available off-the-shelf, to check the properties of some form, or simply to spark some ideas. The importance of the supplier is also reflected in the results presented by Bottle and O'Connor [62] who conducted a survey of the information seeking patterns of designers, in which they highlighted the sources of information most commonly used frequently:

79 %: Consulting personal files, books and papers.
71 %: Contacting a manufacturer directly.
69 %: Manufacturer's catalogues.
65 %: Contacting colleagues.

Similarly, research by the authors involving a detailed questionnaire survey of over 200 designers [28] has shown that 91 % of respondents used supplier catalogues, and more significantly obtained a wide variety of information from this source such as application examples, installation techniques, life and load calculations and so on. These information types have broadly been classified into the following four areas:

- *Commercial Knowledge* - information specifically concerning costs, delivery times and other commercial factors.
- *Overview Knowledge* - information of a global nature that gives an overview of a particular product area, such as names and addresses of key suppliers, types of product available and so on.

- *Soft Product Knowledge* - information giving more detailed product data such as design principles, overall functional and performance characteristics, and design trends and limitations within a particular domain.
- *Hard Product Knowledge* - this comprises the detailed information which gives specific product dimensions, performance details, reliability values and the like or details of manufacturing process capabilities and production system capacity.

An additional unexpected finding in this work was the extensive use of suppliers design guides for specific analytical or procedural design activity. 80 % of the respondents used them, and of these, 50 % used them at least frequently.

3.2 Suppliers in New Product Development

The above has predominantly focused on suppliers as an information source in relation to the provision of standard components, but in an increasingly competitive world market, ultimate success will depend upon the ability of an organisation to integrate a wide range of bespoke technologies into its products. This in turn will necessitate increasing involvement of external suppliers and sub-contractors. However, as a large part of the expertise in engineering is distributed throughout the design and manufacturing functions of organisations that collaborate in the design and manufacture of a product [60], suppliers will increasingly become an integral part of the design team, where they can directly provide design experience for their own products application to the overall products' design. For example, in a recent report by the UK Department of Trade and Industry (DTI) it is suggested that greater out-sourcing of products will be required and based upon more rational and co-operative relationships between manufacturers and suppliers, which to quote *"... include : joint technological development and full involvement of suppliers in the design process ..."* [63].

In this context, the importance of the customer-supplier relationship is now recognised. However, this has been further complicated by the introduction of Concurrent Engineering (CE) techniques, the implementation of which calls for the requirement of a multi-discipline team utilising functions such as: design; production; specialists (e.g. materials and stress engineers); purchasing; management; and more importantly the customer, where the emphasis of 'design teams' has been strongly advocated as the key to successful new product development [64]. CE teams, by their very nature, focus on the product to the total organisation (product values, customer wants and enterprise interests). These CE teams are able to design world-class products and deliver to market in a shorter time span that their competitors. The best of such successful enterprises have found that people working in cross-functional teams and adopting CE principles are less likely to make mistakes [64].

With the adoption of CE the design process is undertaken in parallel or as multitasks, which results in the second stage commencing before the first one is complete and so on. Hence, one area of concern for enterprises adopting the CE approach, is that engineering designers now have to work with less detailed or

partial information, with incomplete knowledge and subjective interpretations. Because much of this information is incomplete, communication between the design team members is much more important. They must be able to transfer information quickly and in both directions. For example, the engineering designer must be able to provide information to the supplier as well as continually updating the product design specification. Similarly, the supplier must be able to provide information to the engineering designer as well as to production planners. This flexibility also enables the design team to provide feedback on how well the design meets their current needs during each phase of the product's development.
A significant amount of work has been undertaken, in the general area of design and manufacturing management, in which the role of the supplier is stressed, but in which there is little in the way of detailed guidance on how suppliers may be successfully integrated into the CE design process. The DTI/PA Consultants report on 'Manufacturing into the late 1990s' highlights in considerable detail the role of the supplier, categorises supplier type and highlights how important is the relationship [63]. In particular, it is quoted that "...as a product becomes more multi-technology external sourcing will increase..". However, no guidelines, procedures or techniques are cited to assist the engineering designers in the integration activity. Vonderembse and White [65], in their book on operations management, have a chapter on 'New product development: a team approach' in which they include a matrix of involvement during the product development cycle. This actually categories the role of the supplier to "...input to the design of the product or process...", yet they also omit to give any detail of how this may be achieved. Cunningham and Homse [66], without explicitly referring to the designer, focus on the role of personal contacts and Martin [67] states that "... increasingly it is information that makes the difference between a marginal supplier and a good one...". Hollins and Pugh [68], in their work on successful product design, imply that the extensive use of external supplier input is a key delineation between dynamic and static design.

3.3 Customer-Supplier Relationships in Concurrent Engineering

Many enterprises have now adopted CE practices, an influential factor for which has been the location or proximity and the relationships of the design team members. The closer that members of the design team are, the better and more enhanced the communication [69], enabling the team members to assimilate the viewpoints of others, incorporating them into their own understanding of the product. Smith and Reinertsen [70] discuss this at length and state *"A great deal of communication occurs in collocated teams simply from the ease of hearing another person's conversation directly, ..."* and *"When all of the team's members are located close enough to each other to be exposed to what their colleagues are saying, much of their colleagues' information becomes theirs automatically"*. Within his research Allen [69] considers the relationship between the physical separation of team members and their subsequent communications, by establishing that there is a link between the distance of separation and the probability of communication. The research also highlights that in order for

members of the team to communicate effectively, they must be located closer than ten metres in proximity to each other and that beyond thirty metres collocation provides little or no benefit and they might as well be located across the city.
These issues have an obvious impact on the relationships between a customer and its suppliers during new product development. For example, in their widely respected book "The Machine that Changed the World", Womack *et al.* [71] have cited examples of major component and sub-system suppliers locating or being forced to locate in close proximity to a customer's major factories. This however would only be feasible for very large contracts, and even then it may not be practical in all instances. Hence in the majority of multi-organisational product design situations, geographic distance will undoubtedly form a barrier. This problem however can be minimised by the use of electronic communication [11] [72], such as: fax; e-mail; video conferencing; electronic data interchange (EDI); etc., ensuring that information is passed on and received almost as quickly as if they were collocated. A fax is a form of electronic communication, but it is basically a fast way of sending a letter whereas EDI is the automated exchange of business information between organisations [54]. Paper elimination, as shown in Figure 2, is only a minor advantage of EDI.

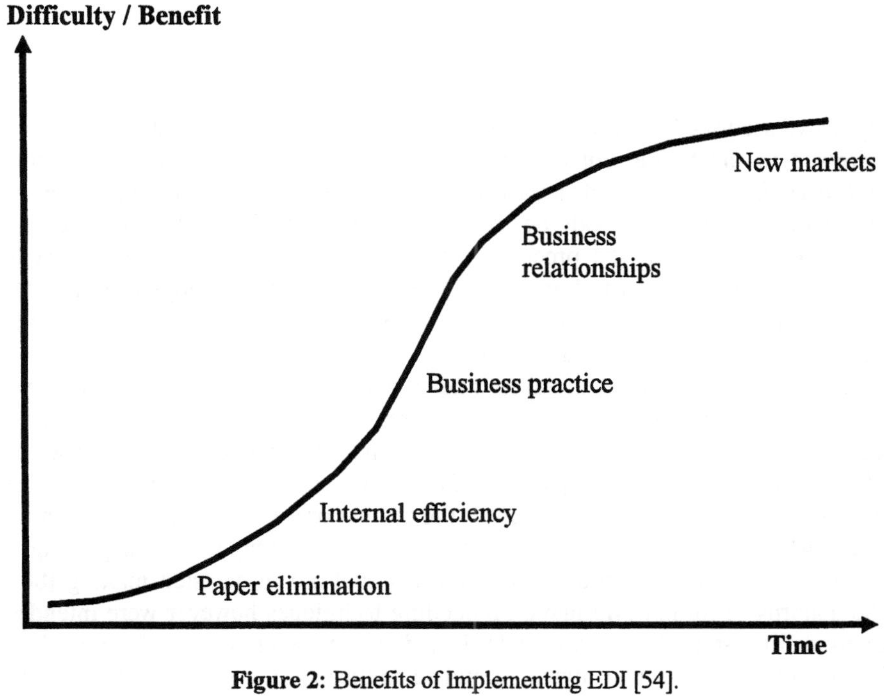

Figure 2: Benefits of Implementing EDI [54].

The substantial benefits of EDI are achieved through an increased internal efficiency, more efficient business practices and long term partnerships [54]. The use of EDI to better co-ordinate the actions of a customer and supplier is often cited as a mechanism by which costs could be reduced or new markets accessed [73].

EDI offers a major improvement by speeding up the flow of information and hence the flow of goods and money. The benefits cited above are of obvious advantage owing to the astronomical number of information exchanges during product development [25], and sooner or later all companies will be forced to use EDI, as customers will impose it on their suppliers, who in turn will do the same with their suppliers.

Such media for communication have been observed in practice by Harrison and Minneman [74] when studying engineering designers engaged in the design activity. They conclude that the need for designers to communicate is increasing as projects become more complex and design teams become more distributed. They state that *"... emerging communication solutions available to designers will have profound effects on the way design is practised ..."* and more importantly, that *"... the design research community must actively participate in the design of such systems ..."*. However, it is interesting to note that more recently Hameri and Nihtilä [72], who studied several hundred participants using electronic communication (WWW) in new product development, reported that periodic face-to-face communication is a necessary prerequisite for all kinds of collaboration, be it electronic or not.

3.4 Preliminary Studies

The above sections have given an insight into some of the issues that need to be addressed in the future, but before developing information systems to support multi-organisational product design and manufacture, there is a need to better understand and document both the activities of the people [75] and the form of the data which is used in the process [10] [76] [77] [78]. This is particularly important as information technology (IT) links alone between organisations will not generate the claimed benefits unless a restructuring of the nature of work within organisations is undertaken [79].

Hence, the authors have undertaken studies pertaining to the information flows and relationships between customers and suppliers engaged in multi-organisational concurrent product design and manufacture. The use of information modelling techniques for this research has been advocated [20]. The modelling process not only helps clarify the current process but also helps identify places where the process can be improved, whereby graphically displaying a problem or coupling can provide a powerful visual aid in decision making [80]. The majority of formal information modelling techniques however were intended for modelling static business data [15], and this presents a major barrier since design information is both dynamic and semantically rich [81]. Moreover, the models need to represent the communication media employed, the types, formats and quantities of information [82], and the impact that this information has on the concurrent engineering design process, but this is not practical with existing techniques. Hence, a new modelling technique to address these requirements was developed by the authors [2] and this has subsequently been implemented in a software application. It is a dynamic modelling tool which can be used in 'real-time' as a deign project proceeds, and hence it can also perform the role of a project information management tool. Post analysis of a number of completed models is currently being performed, and this has, amongst other things, highlighted inconsistencies in information transfer patterns in very similar customer-supplier interaction scenarios. In particular, it has been noted that a correlation may exist between the designers' perceived domain knowledge and the level of supplier interaction, as shown below in Figure 3. Clearly this is an area of concern, as in the absence of supplier interaction the engineering designers may not be aware of the 'state-of-the-art', and hence products may be of less than optimal quality.

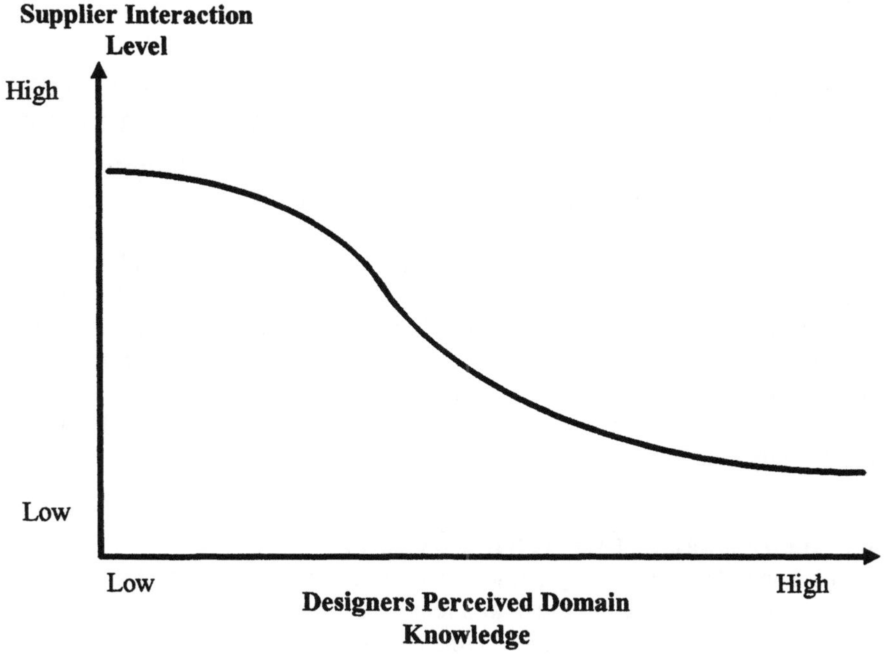

Figure 3 : Supplier Interaction v Designers Knowledge

Similar findings have been reported by Devine and Kozlowski [83] in the context of decision accuracy, where they found that high knowledge individuals (who are significantly more accurate than low knowledge individuals on well structured decisions) reduced their search for information when decisions were well structured. In the authors' domain of study, it is believed that these findings can be attributed to a lack of formal guidelines for supplier integration. This is hardly surprising considering that the majority of the published literature relating to suppliers in engineering has been focused at a much higher level of abstraction, concentrating on for example supply chain management and relationship building, with no specific reference to the content and process of exchange during the design process [84].

3.5 Summary

The relationship between engineering designers and the companies that supply them with products, materials, manufacturing and finishing processes is a highly complex one. It may exist at a direct and frequent contact level at one end of the spectrum, and may not exist at all at the other end. The key element to this relationship is the way that information of both a formal and informal nature flows between the participants. This section of the paper has provided an insight into some of the issues that are involved and presented some brief findings from modelling customer-supplier interactions. As a result of this research, a new improved understanding of the way that information is shared and the nature of the interactions between engineering designers and suppliers is beginning to emerge.

4. THE USE OF PERSONAL KNOWLEDGE AND EXPERIENCE IN NEW PRODUCT DEVELOPMENT

The above discussion has focused on the importance of information to the development of new products and the issues involved in its representation and subsequent use. One common theme observed was the use and reliance made upon the individual designers knowledge and experience. In fact, Fleischer and Liker [30] claim that the engineer is often the main repository of institutional memory about parts and components, which must be re-applied in the future. This has further ramifications, as if it is concentrated within a few individuals it may be lost, either through retirement or job rotation [36].

This section will concentrate on the importance and use of personal knowledge and memory held by the individual engineering designer.

4.1 Knowledge in Engineering Design

In his book, Ullman [55] proposes that three types of knowledge exist that engineering designers make use of and refer to during their work :

- *General Knowledge*, gained through everyday experiences and general education. The information used in updating this knowledge is that which most people know and apply without regard to the specific domain that they are working in.
- *Domain-Specific Knowledge*, gained through study and experience within the specific domain that the designer works in. Information is on the form or function of individual items or groups of items.
- *Procedural Knowledge*, gained from experience of how to undertake ones tasks within the enterprise concerned. This form of knowledge is often based upon a combination of the previous two.

Similarly, Ferguson [85] considers engineering knowledge in this manner, with the added emphasis of the domain in which the engineer works. He states that *"The formal knowledge that engineering designers use is not science, although a substantial part of it is derived from science. It includes as well knowledge based on experimental evidence and on empirical observations of materials and systems"*. This would suggest that knowledge can only be gained by engineering designers having a good appreciation of the area and the environment in which they work.

This view is also presented by Vincenti [86] in his book 'What Engineers Know and How They Know It'; through his observations of the historical developments in aeronautical engineering. He argues that knowledge is developed and formalised to meet the needs of engineering designers in a particular domain, and that some items of this knowledge are clearly distinguishable, whilst others are not. Vincenti proposes six *knowledge categories* that all engineering designers should possess or have at least have access to :

- Fundamental design concepts.
- Criteria and specifications.
- Theoretical tools.
- Quantitative data.
- Practical considerations.
- Design instrumentality's.

Eder [35] further categorises knowledge as either *prescriptive* or *descriptive knowledge*. The former is that commonly referred to as the 'know-how' of design and includes :

- Design knowledge related to the technical system to be designed (knowledge about natural phenomena, knowledge about how to apply that science, etc.).
- Design knowledge related to the design process (knowledge about general strategic approach to designing, knowledge about tactics and methods for designing, etc.).

and the latter the 'know-that' of design and includes :

- Design knowledge related to the technical system to be designed (knowledge about properties and constituents of socio-technical and technical systems, knowledge of theories of properties).
- Design related to the design process (knowledge about design processes, knowledge about using working means).

However, this distinction between know-how and know-that has been criticised by Lera *et al.* [87] when referring to the education of architectural designers. They state that *"It is our view that specialists may be putting undue emphasis on teaching the 'know-that' information of their discipline, rather than attempting to provide an educational context which will promote the architectural design skill of 'know-how'"*. The main emphasis of their research was to ensure that students are made aware of the body of knowledge and that they are able to integrate it into their design processes and problems. The same can be said for engineering designers. Many recent graduates have an in-depth knowledge of facts and figures, but are unfamiliar with the methods of applying them to real problems; such methods requiring a long period of training to establish.

In a similar approach to that described above, the authors have also considered information in the form of knowledge [88], by classifying it into two categories:

- *General Knowledge* is that applicable to say a component regardless of the details of its design, its purpose and the person designing it. It is usually a well established fundamental principle based on performance or size (e.g. the equations for a gear system).

- *Expert Knowledge* is that used for designing to greater detail, and depends on the purpose and the expert carrying out the design. This can take many forms, such as: designer knowledge (heuristics, rules of thumb); component knowledge (standards, catalogues) and company specific data (standards, requirements).

4.2 Information and Memory in Engineering Design.

All engineering designers at some stage in their work, rely on the use of information stored within their own memory; whether this is a specific detail of for example a material strength, simply information of where certain things are stored or located, or information on how to obtain things. Of equal importance to this is the link between knowledge and memory, owing to the fact that knowledge is contained within the individual's memory. However, the working memory is not well understood, as there are no methods or ways of directly measuring:

- What is held in an individual's memory.
- How information is accessed from this memory store.
- How much memory the individual possesses.

Nevertheless, it is the association between the information received through ones senses and the knowledge stored in ones memory that is of interest here. This association produces actions and responses from the individual when receiving and passing on information. However, different individuals respond differently to the same information even if the task they are undertaking is the same.

Research into the use of memory by individuals has been widespread and predominantly aimed towards the problem-solving methods and skills of the individual, which branches into the domain of cognitive psychology [23] [55] [89] [90] [91]. Such research has concentrated on the investigation into the characteristics of the activities of the human mind. Emanating from this research are models that give an inclination into our memory during design activities. The memory is generally considered to consist of two different types: *short term memory (STM)* and *long term memory (LTM)*.

- STM was described by Ullman [55] as the main processor in the human brain; which is analogous to the operating memory of the computer (more commonly referred to as its random access memory (RAM)), but of limited capacity.
- LTM is comparable to a computing analogy in a similar manner to that of STM. LTM is analogous to the computer's disk storage, which is for permanent retention and therefore refers to all of the information held within the memory of the individual, which has been accumulated over a long period of time.

4.3 The Different Levels of Memory

Empirical research by the authors has shown that engineering designers make extensive use of their memory to aid in the development of new products [6]. It was observed that in many cases engineering designers rely solely on recalling items of information and data from their memory rather than spending a large amount of time searching for it. The research has shown that this memory is used for a number of different purposes, which have been categorised in four ways :

- To retrieve **Direct Data** held within the engineering designers memory.
- To retrieve **Knowledge** of the source of the information or data from memory.
- The use of **Vague Memory** to retrieve the rough location of where the information or data is stored or may be found.
- To retrieve details of the **Process** required to undertake a design activity.

The significance of the use of an engineering designers memory under these categories is discussed in the following section.

4.4 The Significance of Memory Usage.

Observations made by the authors from 20 case-studies of engineering designers undertaking a number of design activities has shown that the engineering designers previous experience and knowledge, and in particular **memory**, is of vital importance to the development of a product [92]. It was therefore decided to investigate this aspect of the research in more detail. During a case-study, each information access that was made to **memory** was analysed against the raw data retrieved during the application of an extended IDEF1X modelling technique, using the four categories identified above. Table 1 below shows further analysis of the widespread use of these different categories of memory usage for the 20 case-studies.

The importance of memory usage is clearly seen by comparing the total number of information accesses to that proportion retrieved from memory. It is seen that almost one third of the overall total information accesses (30%) are based upon memory usage; with many of the individual case-studies being more than this, for example case-study No's: 6, 18 and 20 are based upon over 50% memory usage, and case-study No's: 1, 4, 11 and 17 over 40%. These are significant research findings, in that they highlight an area of information accessing that has not been widely considered by design researchers in general. A number of design researchers have however, considered knowledge usage by the engineering designer. Research by Stauffer and Ullman [23], identified *knowledge* (formed within memory) to be of great importance to the engineering designer. They observed that *"The essence of mechanical design, seems to rely very little on problem solving techniques but rather on a richness of knowledge"*. Similarly, Lewis [93] has identified this to be important, but in terms of the engineering designers experience. He states that *"The exercise of judgement is to be distinguished from guessing or playing hunches in that there is a conscious referral back to previous experience"*. However, this research has quantified the actual usage of knowledge and in particular the use of **memory** within a wide range of design activities.

Case Study No	Direct Data	Knowledge	Vague Memory	Process	Total Memory Accesses	Total No of Information Accesses	Memory Access as a % of Total
1	2	4	1	0	7	17	41%
2	1	0	0	0	1	9	11%
3	0	3	1	0	4	16	25%
4	2	2	0	0	4	10	40%
5	1	1	0	0	2	11	18%
6	3	3	1	1	8	13	62%
7	1	2	0	1	4	11	36%
8	0	2	0	0	2	7	29%
9	2	2	0	0	4	12	33%
10	0	1	1	2	4	17	23%
11	3	3	0	1	7	17	41%
12	1	1	0	0	2	18	11%
13	0	0	0	0	0	14	-----
14	0	0	0	0	0	5	-----
15	2	0	0	1	3	17	18%
16	0	0	0	1	1	13	8%
17	5	3	0	0	8	17	47%
18	4	0	1	1	6	11	55%
19	2	2	0	0	4	11	36%
20	1	3	0	0	4	7	57%
Total	30	32	5	8	75	253	30%

Table 1 : Breakdown of engineering design memory usage.

The findings from this research have obvious implications for the provision of future information systems for engineering designers.

5. CONCLUSIONS / FUTURE RESEARCH ISSUES

As the millennium approaches, the role of information for business organisations in general, will become more and more important, especially in the context of engineering design. This is because of its role in assisting companies to maintain their competitive edge, both in terms of core technologies and manufacturing techniques.

The diversity and explosive pace of developments in the fields of: materials; components; analytical methods; CAE strategies; control and systems techniques; new legislation; etc., as seen in the research and technical press and at any trade show or exhibition, will have to be handled in a more structured and organised manner than at present. Similar developments are taking place in the field of manufacturing where new net shape methods, machining techniques, assembly strategies, etc., are being developed on a regular basis.

The importance of legislation and externally imposed standards will also impose an increasing burden. Companies will need to show and even prove that their products have been designed using appropriate and proper techniques, both to obtain business and to avoid litigation.

Thus this paper has described key considerations in the domain of design information and in particular has shown the key role of suppliers and the extensive use of the individuals' knowledge within the design process. It is clear that it may not be possible for individual engineering designers to improve their awareness of new technologies, techniques and legislation using conventional educational means, particularly as companies place such a low priority on updating and training their staff [94]. There is thus a need for the development of designer specific information systems, some key elements for such are suggested below.

5.1 The Specification for Future Designer Specific Information Systems

The preliminary guidelines for such a system are listed below, but as a prerequisite it would be necessary to develop a simple classification and categorisation schema to support the types of occurrences listed below:

- The ability to record the personal knowledge of the engineering designer.
- A system that records the historical design decisions made about the enterprises products.
- The ability to record the engineering designers memory.
- A system that acts as a memory jogger/prompter of where information is located or stored.
- The ability to store 'soft' and very unstructured information (notes, comments, observations etc.).
- A system that provides access to current supplier information.
- The ability to interact with suppliers and subcontractors.

Some progress is being made in these areas, but considerably more research and development is required before this level of support is available on every designer's desk.

Acknowledgements and References

The authors wish to thank the Engineering and Physical Sciences Research Council (EPSRC) from which the research work was funded, under Grant Nos. GR/G62387 and GR/K59613.

1. Benyon D 1990 *Information and Data Modelling*. Alfred Waller Ltd, Henley-on-Thames, UK.

2. Boston O P, Culley S J, McMahon C A 1996 Designers and Suppliers - Modelling the Flow of Information. *Proceedings of ILCE '96*, Paris.

3. Cave P R, Noble C E I 1986 Engineering Design Data Management. In: Leech D J, Middleton J, Pandle GN, (eds.), *Proceedings 1st International Conference on*

Engineering Management : Theory and Applications, Swansea UK, M. Jackson & Son, pp 301-307.

4. Court A W, Culley S J, McMahon C A 1993 The Information Requirements of Engineering Designers. *Proceedings of the Ninth International Conference on Engineering Design*, The Hague, Netherlands, pp 1708-1715.
5. Putre M 1991 Product Data Management. *Mechanical Engineering*, October, pp 81-83.
6. Court A W, Culley S J, McMahon C A 1996 Information Access Diagrams: A Technique for Analysing the Usage of Design Information. *Journal of Engineering Design*, Vol.7, No.1, pp 55-75.
7. Shannon C E, Weaver W 1949 *The Mathematical Theory of Communication.* University of Illinois Press, USA.
8. Pahl G, Beitz W 1984 *Engineering Design.* Wallace K M (ed), The Design Council, London, UK.
9. Rouse W B 1986 On the Value of Information in System Design: A Framework for Understanding and Aiding Designers. *Information Processing and Management*, Vol.22, pp 217-228.
10. Rzevski G 1985 On Criteria for Assessing an Information Theory. *Computer Journal*, Vol.28, No.3, pp 200-202.
11. Bush J B, Frohman A L 1991 Communication in a "Network" Organization. *Organizational Dynamics*, Vol.20, No.2, pp 23-36.
12. Daft R L, Lengel R H 1984 Information Richness: A New Approach to Managerial Behaviour and Organisation Design. *Research in Organisational Behaviour*, Vol.6, pp 191-233.
13. Holland W E, Stead B A, Leibrock R C 1976 Information Channel/Source Selection as a Correlate of Technical Uncertainty in a Research and Development Organization. *IEEE Transactions on Engineering Management*, Vol.23, pp 163-167.
14. Liker J, Hancock W 1986 Organizational Systems Barriers to Engineering Effectiveness. *IEEE Transactions on Engineering Management*, Vol.EM-33, pp 82-91.
15. Rangan R M, Fulton R E 1991 A Data Management Strategy to Control Design and Manufacturing Information. *Engineering With Computers*, Vol.7, pp 63-78.
16. Turner B T 1977 The Best Format for Design Information. *Information Systems for Designers*, University of Southampton, UK, pp 87-94.
17. CALS 1989 *Application of Concurrent Engineering To Mechanical Systems Design.* CALS TR 002, Washington, DC, USA.
18. STEP 1988 *Standard for the Exchange of Product Model Data.* ISO TC184/SC4/WG1, October, DP 10303.
19. Sanvido V E, Kumara S, Ham I 1989 A Top-Down Approach to Integrating the Building Process. *Engineering With Computers*, Vol.5, pp 91-103.
20. Chadha B, Fulton R E, Calhoun J C 1991 Case Study Approach for Information-Integration of Material Handling. *ASME Engineering Databases: An Engineering Resource.*

21. Poon S Y 1991 Information Systems - A Management Question. *Professional Engineering*, January 1991.

22. Botterill E 1992 Cutting the Paper Mountain. *Professional Engineering*, January 1992.

23. Stauffer L A, Ullman D G 1991 Fundamental Process of Mechanical Designers Based on Empirical Data. *Journal of Engineering Design*, Vol.2, No.2, pp 113-125.

24. Macleod A, McGregor D R, Hutton G H 1994 Accessing of Information for Engineering Design. *Design Studies*, Vol.15, No.3, July, pp 260-269.

25. Safoutin M J, Thurston D L 1993 A Communications-Based Technique for Interdisciplinary Design Team Management. *IEE Transactions on Engineering Management*, Vol.40, No.4, November, pp 360-372.

26. Shuchman H L 1981 *Information Transfer in Engineering*. Glastonbury, Coun.: The Futures Group, ISBN 0960519602.

27. Wilkin A 1981 *The Information Needs of Practitioners : A Review of the Literature*. University of London, British Library R&D Report No 5611, February.

28. Court A W, Culley S J, McMahon C A 1994 Information Sources and Storage Methods for Engineering Data. *Proceedings of the 2nd ASME Biennial European Joint Conference on Engineering Systems Design & Analysis*, PD-Vol.64-5, pp 9-16.

29. Schwarzwalder B 1992 Information Access: More Options for Engineers. *Automotive Engineering*, February 1992.

30. Fleischer M, Liker J K 1992 The Hidden Professionals: Product Designers and Their Impact on Quality. *IEE Transactions on Engineering Management*, Vol.39, No.3, August, pp 254-264.

31. Pitts G 1983 Retrieving and Handling Design Information. *Efficiency in the Design Office*, Mechanical Engineering Publications, UK, pp 49-53.

32. Raman P G 1974 A Structure for Design-Related Information. *Information Systems for Designers*, University of Southampton, Ch.4.

33. Yeaple R N 1992 Why are Small R & D Organizations More Productive. *IEEE Transactions on Engineering Management*, Vol.39, No.4, November, pp 332-346.

34. Fechter W F 1993 Competitive Myth. *Quality Progress*, No.5, pp 87-89.

35. Eder W E 1989 Information Systems for Designers. *Proceedings of the International Conference on Engineering Design*, pp 1307-1319.

36. Salzberg S, Watkins M 1990 Managing Information for Concurrent Engineering: Challenges and Barriers. *Research in Engineering Design*, Vol.2, pp 35-52.

37. Levy S, Subrahmanian E, Konda S, Coyne R, Westerberg A, Reich Y 1993, *An Overview of the n-dim Environment*. Engineering Design Research Centre, Report No. EDRC -05-65-93, January.

38. Culley S J, Webber S 1992 Implementation Requirements for Electronic Standard Component Catalogues. *Proceedings of IMechE Journal of Engineering Manufacture*, Vol. 296, 1992.

39. Armour J 1993 A New Chapter Opens for Trade Catalogues. *Information Technology*, June 1993.

40. Will P 1991 Simulation and Modelling in Early Concept Design. *International Conference on the Application of Manufacturing Technologies (ICAMT)*, Alexandria, VA, April 18th.

41. Schierbeek B B 1989 The Profile of the Engineering Designer in the Nineties. *Proceedings of the International Conference on Engineering Design*, pp 201-210.

42. Bradley S R, Agogino A M 1990 Knowledge Capture for Concurrent Design. *Proceedings of the Winter Annual Meeting ASME Prod Eng. Div.*, Vol.47, pp 17-30.

43. Reynard K W 1991 Extracting Metals Information. *Professional Engineering*, June 1991.

44. MacCallum M 1986 Computer Algebra : Tomorrow's Calculator. *New Scientist*, No.1531, pp 52-55.

45. Stauffer L A, Ullman D G, Dietterich T G 1987 Protocol Analysis of Mechanical Engineering Design. *Proceedings of the International Conference on Engineering Design*, pp 74-85.

46. Kuffner T A, Ullman D G 1991 The Information Requests of Mechanical Design Engineers. *Design Studies*, Vol.12, No.1, January, pp 42-50.

47. Almli F 1988 A Study of Information Flow During the Product Development Process. *Proceedings of the International Conference on Engineering Design*, pp 17-23.

48. Wong A, Sirram D 1993 SHARED: An Information Model for Co-operative Product Development. *Research in Engineering Design*, Vol.5, pp 21-39.

49. Rasmussen J 1985 Value of Information in Supervisory Control. *Proceedings of the Second Symposium on Empirical Foundations of Information and Software Science*, Georgia Institute of Technology, Atlanta, October.

50. March J, Trott F 1988 Databases for Engineers. *Professional Engineering*, Dec 1988.

51. Nochur K S, Allen T J 1992 Do Nominated Boundary Spanners Become Effective Technology Gatekeepers ?. *IEE Transactions on Engineering Management*, Vol.39, No.3, August, pp 265-269.

52. Sheen R 1992 Barriers to Scientific and Technical Knowledge Acquisition in Industrial R&D. *R&D Management*, Vol.22, No.2, pp 135-143.

53. Clark K B 1989 Project Scope and Project Performance : The Effect of Parts Strategy and Supplier Involvement on Product Development. *Management Science*, Vol.35, No.10, pp 1247-1263.

54. Blotwijk M 1993 EDI in the Process Industry. *Human Systems Management*, Vol.12, pp 49-53.

55. Ullman D G 1992 *The Mechanical Design Process*. McGraw-Hill, Inc.

56. Sanford L 1989 Managers Need Total Information. *Professional Engineering*, October 1989.

57. Bögler D 1995 Computers on the Blink. *The Daily Telegraph*, September 18th 1995.

58. Granstrand O, Bohlin E, Oskarsson C, Sjoberg N 1992 External Technology Acquisition in Large Multi-Technology Corporations. *R & D Management*, Vol.22, No.2, pp 111-133.

59. Allen T J 1969 The Differential Performance of Information Channels in the Transfer of Technology. In: Grubber W H, Marquis D G (eds.), *Factors in the Transfer of Technology*, MIT Press, Cambridge, MA, USA.

60. Bond A H, Ricci R J 1992 Co-operation in Aircraft Design. *Research in Engineering Design*, Vol. 4, pp 115-130.

61. Radcliffe D F, Lee T Y 1991 Observing Engineering Designers at Work. *International Mechanical Engineering Congress*, Sydney, July, pp 4-8.

62. Bottle R T, O'Connor M R 1979 Information Seeking Patterns for Industrial Designers. *Information Systems for Designers*, University of Southampton, pp 23-27.

63. HMSO 1989 *Manufacturing Into the Late 1990's*. A Report by PA Consultants for the Department of Trade and Industry (DTI), Her Majesty's Stationary Office (HMSO), London, UK.

64. Nevins J L, Whitney D *1989 Concurrent Design of Products and Processes : A Strategy for the Next Generation.* McGraw-Hill Publishing Company, USA.

65. Vonderembse M, White G P 1991 *Operations Management: Concepts, Methods and Strategies*. West Publishing Company.

66. Cunningham M T, Homse E 1984 The Roles of Personal Contacts in Supplier-Customer Relationships. *UMIST*, UK, Occasional paper No 8410.

67. Martin J M 1989 Making Information Flow *Manufacturing Engineering*. May, pp 75-78.

68. Hollins W, Pugh S 1990 *Successful Product Design* Butterworths, UK.

69. Allen T J 1977 *Managing the Flow of Information.* MIT Press, Cambridge, MA, USA.

70. Smith P G, Reinertsen DG 1991 *Developing Products in Half the Time.* Van Nostrand Reinhold.

71. Womack J P, Jones D T, Roos D 1990 *The Machine that Changed the World.* Rawson Associates, NY, USA.

72. Hameri A P, Nihtilä J 1995 Distributed New Product Development Based on Electronic Communication - A Case Study. *Journal of Product Innovation Management*, Helsinki, March, pp 1-11.

73. Henderson J C *1989 Building and Sustaining Partnerships Between Line and I/S Managers.* Massachusetts Institute of Technology, Centre for Information Systems Research, Report No.195, September.

74. Harrison S R, Minneman S L 1993 Tools, Communication, and the Nature of Design. *Proceedings of the International Conference on Engineering Design*, pp 351-354.

75. Galliers R D 1985 An Approach to Information Needs Analysis. *Human-Computer Interaction - INTERACT'84*, pp 619-628.

76. Shaw N K, Bloor S, Pennington A de 1989 Product Data Models. *Research in Engineering Design*, Vol.1, pp 43-50.

77. Pugh S 1977 The Engineering Designer- His Task and Information Needs. *Information Systems for Designers*, University of Southampton, UK, pp 63-66.

78. Stamper R K 1985 Towards a Theory of Information. *Computer Journal*, Vol.28, No.3, pp 195-199.

79. Konsynski B R, Warbelow A 1989 *Co-operating to Compete.* Harvard University Working Paper No.89-02.

80. Pracht W E 1986 Gismo; A Visual Problem-Structuring and Organizational Tool. *IEEE Transactions on Systems, Man, and Cybernetics*, Vol.SMC-16, No.2, March/April, pp 265-270.

81. Rabins M, Ardayfio D, Fenves S, Seireg A, Nadler G, Richardson H, Clark H 1986 Design Theory and Methodology- A New Discipline. *ASME Mechanical Engineering*, August, pp 23-27.

82. Christian A D, Seering W P 1995 A Model of Information Exchange in the Design Process. *ASME Design Engineering Conferences*, Vol.83, No.2, pp 323-328.

83. Devine D J, Kozlowski S W J 1995 Domain-Specific Knowledge and Task Characteristics in Decision Making. *Organisational Behaviour and Human Decision Processes*, Vol.64, No.3, pp 294-306.

84. Harland C M 1995 *Networks and Globalisation - A Review of Research.* EPSRC Grant No. GRK 53178, Warwick Business School, University of Warwick, UK.

85. Ferguson E S 1992 *Engineering and the Mind's Eye.* MIT Press, Cambridge, USA.

86. Vincenti W G 1990 *What Engineers Know and How They Know It : Analytical Studies From Aeronautical Engineering.* Baltimore, USA.

87. Lera S, Cooper I, Powell J A 1984 Information and Designers. *Design Studies*, Vol.5, No.2, pp 113-120.

88. Theobold G 1992 *Modelling the Elements of Engineering Sub-Assemblies Based on Standard Components.* Internal Report, University of Bath, UK.

89. Miller G A 1965 The Magical Number Seven, Plus or Minus Two : Some Limits on Our Capacity for Processing Information. *The Psychological Review*, Vol.63, No.2, pp 81-97.

90. Newell A, Simon H A 1972 *Human Problem Solving.* Prentice-Hall, NJ, USA.

91. Simon H A 1969 *The Sciences of the Artificial.* MIT Press, Cambridge MA, USA.

92. Court A W, Culley S J, McMahon C A 1995 Modelling the Information Access Methods of Engineering Designers. *ASME Design Engineering Conferences*, Vol.83, No.2, pp 547-554.

93. Lewis W P 1981 The Role of Intelligence in the Design of Mechanical Components. *Man-Machine Communication in CAD/CAM*, North-Holland Publishing Co, IFIP.

94. Culley S J, Owen G W, Pugh P 1996 *Current Issues in Design - Survey 1995/96.* Published by the IMechE, ISBN 1 85790 027 8.

Improving Design Management in the Building Industry

A N Baldwin, S A Austin and M A P Murray
Loughborough University and AMEC Design & Management

Abstract

This paper describes current approaches to the management of the design process as practised in typical UK building projects and the nature of the problems and challenges faced by this industry. The focus is on management of design information flow between the client (usually the building owner/operator) and the members of the design team (architect, structural engineer, quantity surveyor etc.) and also possibly the contractor. Existing management of design in the building and construction industry is described together with the development of new tools and techniques to assist the members of the design team.

1. Introduction

There is now an increased awareness of the need for better design management in the building industry. This is because of a number of different factors that reflect both changing market conditions and new building materials and processes [1]. A reduced industry workload and tighter fee scales have focused on the need for better management of the design process if profit margins are to be maintained.

More complex buildings have necessitated design input from an increasing range of specialist sub-contractors. New and different forms of procurement have meant that design work, and the deliverables of the design process, must be linked to the letting of the work packages to these sub-contractors. The increasing use of subcontractors has required effective management of the interfaces between these organisations if construction time schedules are not to be interrupted and handover dates delayed [2].

These changes have coincided with industry clients seeking major reductions in the cost of buildings. It is now accepted that, if the construction industry is to produce the cost savings demanded by the Latham report [3] there must be closer integration between the design and construction functions in the product cycle, as has occurred in other engineering sectors (such as the automobile and manufacturing industries). A key aspect is the capability to plan and manage design efficiently, taking into account the changing needs of the client and contractor.

Several different but inter-related approaches to precipitate the required improvements in the design process are currently being evaluated by researchers and industry representatives. These include: improving the briefing process (whereby the client's requirements are identified); using new technologies (such as virtual reality) to provide clients with a clearer understanding of their new building; and closer study of the information flow within the design process.

This paper examines current aspects of design management in more detail. Particular focus is placed on current studies on improving information flow between the parties to the design process. These studies have revealed that the construction industry has recognised the need for better management of the information flow. Poor information transfer results in delayed, abortive or repeated design work. This in turn increases design costs, compromises fulfilment of the brief and reduces construction efficiency. The authors consider that the tools and techniques described in this paper may help overcome some of these problems.

2.The Building Design And Procurement Process

The origins of today's construction industry lie in the industrial expansion of the nineteenth century, when the designer-led arrangement became the established approach to building [2]. To produce a building to meet their requirements clients need to engage both designers and contractors. The exact relationship between the client, the design organisation and the contractor will depend upon the form of procurement chosen. The most common, and the traditional, form of procurement is where the client engages independent designers to design the building, the construction phase of the contract then being let to a suitable contractor following a tendering process where several contractors are invited to bid for the work.

In this process it is normal for the Architect to have the lead role in both the design and the project management of the building. It is the Architect who is first involved with the production of the design brief prior to the conceptual stage of design and who is usually involved in the selection of the other design organisations, structural designers, building services engineers etc. Whist this traditional form of procurement remains the most common, there are several variations and different forms of procurement that have been developed to suit different contractual situations. These include: Schedule of Rates Contracts; Lump Sum Contracts; Cost Re-imbursement Contracts; Target Cost Contracts; Design and Build Contracts; and various forms of 'Management Contracts' [2,4,5] Their selection is a reflection of the individual client, the project and the allocation of the risks involved.

Of all these forms of contract the Design and Build form of procurement is of particular interest as it provides a single point of responsibility for the client. This type of procurement was originally offered by organisations who could provide both design and construction management expertise to the client. Offering design services across a range of engineering disciplines, these organisations were able develop and co-ordinate *all* the design work and then manage the construction process by employing sub-contractors to complete the construction work. This form of procurement has grown in popularity with building clients. In the UK it currently accounts for some £7.7 billion of construction work and equates to 28% of all construction.[6] This level of importance is set to increase as the government's Private Funding Initiative is introduced.

There is now an increasing recognition that to meet the demands of construction clients and to maximise the benefits of information technology construction must be regarded more as a manufacturing process [7]. It is those organisations currently proficient in design and build work who are the best placed to realise the benefits of these new ways of working. Some of these organisations are currently at the forefront of research into new methods of design management.

The need for such research has been recognised by government funding agencies. The Innovative Manufacturing Initiative, (IMI), which was introduced in 1993 has changed the basis for the funding of academic research in the UK. Three programmes within the initiative have already been instigated. Included within these programmes is that of "Construction as a Manufacturing Process" which aims at gaining a better understanding of client's needs through integrated design and construction, flexible off-site fabrication, and improved site productivity. Research projects currently being funded under the IMI scheme include: 'managing the brief'; 'a clients project definition tool'; 'interactive visualisation through VR'; and 're-engineering the processes of design'. Although the IMI initiative seeks to contribute to the improvement of the whole construction process from conception to operation there is increasing awareness of the importance of the clearer identification of client needs and improved management of design process.

3. Current Planning and Management Techniques

Current practice in the planning and management of the design process is focused on the design deliverables, (the drawings, calculations, Bills of Quantities, and specifications), that comprise the submission to the client. These are identified, agreed with the client, (or the client's representative), and listed at the start of each stage of the design process. The tendency is then to simply plan the design process 'backwards" from the date when these deliverables are due to be released to the client. Typically a 'Master Programme' in the form of a bar chart is produced by the project manager . This includes ' global ' activities, together with 'milestones' or key dates. The master programme is distributed to the team leader of each design discipline. Each design team leader then plans the design work of their team within the overall framework of the master programme relying on the transfer of the information required either informally of formally via drawings and at design review meetings.

At the conceptual and schematic stages of the design the Architect traditionally plays the key role in the management of the design process. Resource identification and allocation is limited to a check on the total number of hours work required to produce the deliverables. (This is primarily for cost control and fee earning reasons.) There is usually little attempt at these early stages to allocate resources to specific design tasks. Sometimes a single person within each discipline is responsible for acquiring the information required by other team members.

At the detailed design stage the design tasks may be more clearly identified and there is greater attention to design planing and the scheduling of resources. Resources are managed in such a way that continuity of work is maintained for each design team member. This may require a member of the design team working on more than one project simultaneously. The focus remains on design deliverables. Typical techniques used by design and build organisations include: IRS (Information Release Schedule); and a technique known as EWS, (Early Warning Systems). The IRS system comprises checklists of the information and drawings that are to submitted at certain dates. There is a separate IRS for each subcontractor package within the construction contract. The EWS technique allows closer monitoring of design activities. A mini-programme is produced for each element of the design work and covers working drawings, fabrication schedules etc. Every bar in the design programme is decomposed onto every design element resulting in separate bars representing decomposed design activities in more detail [8].

Topalian, [9], identified twenty-eight major difficulties within the design process. These have been confirmed by more recent researchers [10,11]. Of these difficulties nine are related primarily to the level and timing of information transfer. Hassan [8], in interviews with designers, confirmed the importance of information transfer and the quality of information to the design process. Quality Assurance, (QA), has improved the quality of information transfer but not eradicated the problems For those organisations that have become accredited under BS 5750 or an equivalent a quality plan is produced by the design leader for the project. This mainly comprises a list of the names of the project team members representing the different parties, the procedure for communication between different parties, the procedure for circulating information and procedures for the issue of design documentation. More than quality assurance procedures are needed to improve the management of the design process.

For large and medium sized construction projects network analysis and critical path methods are the generally accepted technique within the industry for the planning and scheduling of construction work [12,13]. Experience has shown, however, that they are inappropriate for design management because of its ill-defined and iterative nature. Design managers now need equivalent tools to help them plan, manage change and integrate their role with the client, contractor and other parties.

4. New Approaches

Recent research has investigated the application and combination of a number of techniques to help in the planning and management of information flow during the building design process, both at early (concept and scheme) and detailed design stages. This research is summarised by Austin et al [14,15,16]; Baldwin et al [17]; and Newton, [1].

These researchers argue for the use of the following techniques as a basis for improving the management of the design process:

(i) data flow diagrams (DFDs);

(ii) matrix representation and partitioning;

(iii) logic networks; and

(iv) discrete event simulation.

Each of these techniques and their use is summarised below:

4.1 Data Flow Diagrams

Investigations have concentrated on the management of the flow of design information, as the studies within industry have shown that this is a key area where improvements could have a significant impact. This in turn requires the design tasks which make up the process to be defined. Data Flow Diagrams are well suited to this objective because they: are data orientated; do not require the processes themselves to be modelled (they are a black box); can be partitioned to produce a hierarchy of tasks, each being sub-divided as required; and do not require the order of processing to be predefined. The latter is important because of the need to be able to both predict and analyse the inter-related tasks that must be iterated through to obtain a satisfactory solution. This is the prime characteristic of design activity that gives it its unique nature and makes management of it such a challenge (some would argue an impossible one).

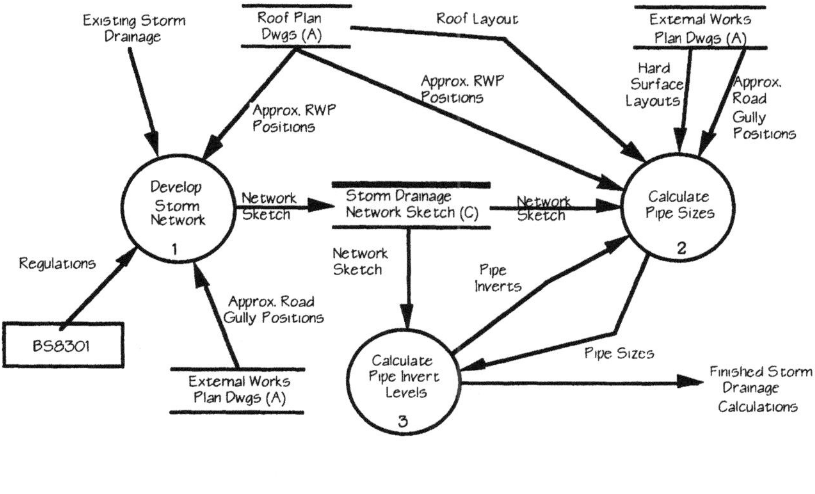

Figure 1 Typical Building-Design Data Flow Diagram, Newton (1996)

Data Flow Diagrams were developed as a Structured Analysis technique for computer programming. Figure 1 shows a typical diagram, which consists of Processes (design tasks), Data flows (design information), Data stores (drawings, calculations, reports etc.) and Sources or Sinks (external bodies to the system, e.g. a regulatory authority). Using a combination of top-down and bottom-up approaches Newton [1] and Hassan [8] have assembled DFDs to represent both the scheme and detailed design phases of building design; these have been termed Design Process Models. This has been achieved through historical analysis of a variety of projects and interviews with designers and team leaders to verify the models.

One of the main objectives of this research has been to identify whether it is possible to work-up a *generic* model for each stage of design. This is considered by Newton and Hassan to be an essential requirement in terms of practical application due to the substantial effort needed to build and verify a model of the building design process. They have found that it is possible to form a generic model, as the individual design tasks and their data flows are independent of the project type. However, this requires the model to have options, in order to deal with a variety of specific solutions (e.g. foundation type: pad, strip, raft and piled). Thus their Design Process Models have optional DFD components, with the appropriate one being selected and the others 'detached'. The hierarchical/optional form of the model is illustrated by Figure 2.

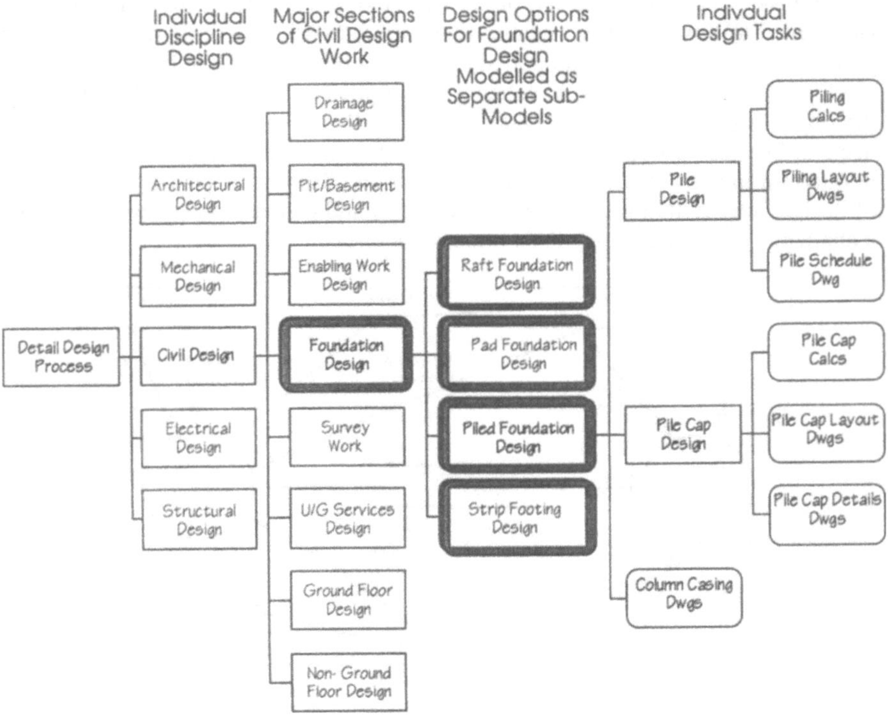

Figure 2. Design Process Model structure

These investigations have also indicated that the Design Process Models can be made independent of the method of procurement. This is an important finding because of the variety of financial, organisational, and delegational arrangements that can be used on a building project. (See section on the building design and procurement process)

However, their work has also shown that it is not possible to use the same Design Process Model for all stages of design. This hypothesis was based on the argument that it might be possible to use the upper (more general) levels of a full model for concept and scheme design and the complete model for detailed design. However, the nature of the processes and information flows was found to be different and hence separate DFD models were produced for each stage. An example of the differences they cite is the design outputs (DFD stores) which mainly consist of reports, sketches and costs at early stages compared to drawings and calculations during the production of complete construction information (detailed design activity). A fuller description of the use of DFDs for detailed building design is reported by Austin et al [15].

4.2 Matrix Analysis

This technique was originally developed by Steward [18] and has been applied by Eppinger et al [19] to production engineering. This has been found it to be a suitable method for analysing the building design process as it can:

(i) identify design tasks that are interdependent (i.e. in loops, not sequential);

(ii) order the tasks to find the optimum in terms of minimizing design iteration;

(iii) identify the natural groupings of tasks (that therefore require careful co-ordination);

(iv) allow the effects of changing the order of design to be analysed and explained; and

(v) are a step towards producing a deign programme via a logic network.

The method consists of a matrix showing all the tasks (denoted by reference numbers) on both axes and a mark to indicated an information need (Figure 3). Marks below the diagonal indicate that the information is available (as the producing tasks has already been carried out), whereas those above show that a 'guess/estimate' must be made (as the producing task has yet to be started). An algorithm is used to sort the tasks into the optimum order (i.e. to get marks close to or ideally below the diagonal) - this is termed partitioning.

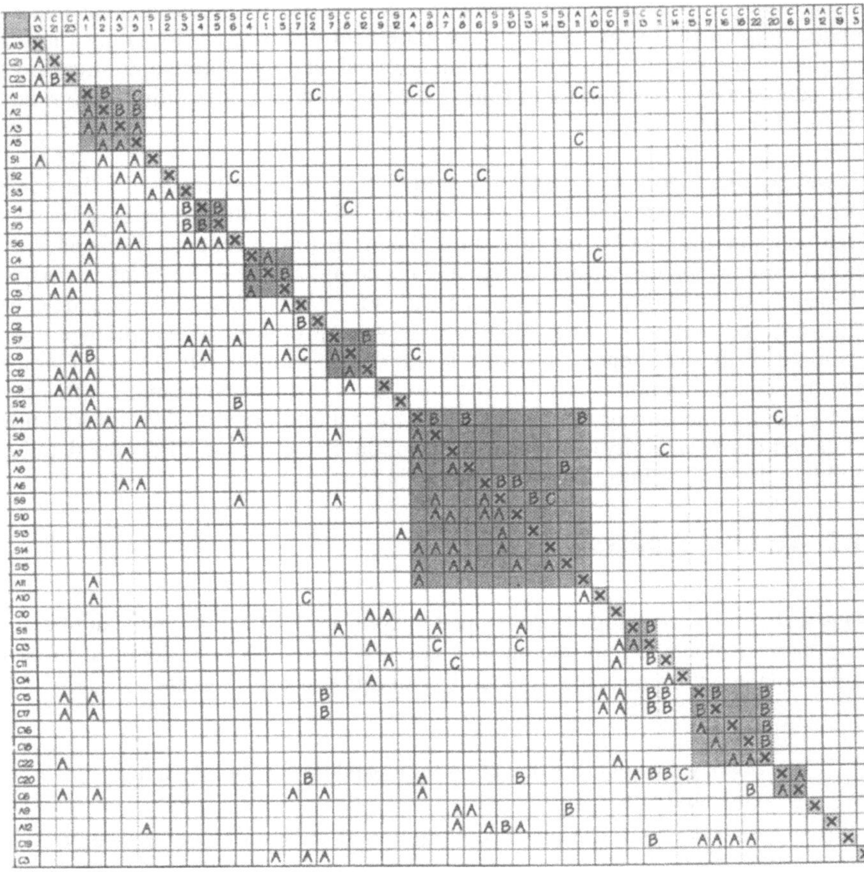

Figure 3. Example of a re-partitioned matrix, Austin et al (1996)

In order to get a sensible analysis it is also necessary to classify the information dependencies (e.g. A = strong .... C = weak). If weak dependencies (e.g. ones where a good estimate can be made) are ignored the matrix can be re-partitioned to produce a suitable design order (Figure 3). An examination reveals that the matrix has identified blocks of interdependent tasks that must be managed with care. Moreover, experienced engineers have agreed that these blocks represent the groupings that are typical of those that blight design; they tend to be at interfaces, e.g. interaction of cladding, glazing, services and structure, and are multi-disciplinary. The latter observation is perhaps obvious but important because it demonstrates that the natural groupings of design activities that must be managed are multi-disciplinary, which is in contrast to current management methods where design is largely planned by each discipline separately.

Another important feature of this matrix technique is that tasks can be moved up or down the matrix, which is then re-partitioned, to allow the effects of change to be analysed. Thus a civil engineer would prefer to design the foundations last, but the contractor would like the information early; the effects of imposing this order on the design can be assessed and costed (including the need to over-design to avoid the possibility of re-design/construction later). Alternatively a client may wish to delay/change a decision which has detrimental down-stream effects on the design team's work.

4.3 Logic Networks And Design Programmes

The next stage in planning is to produce a programme or schedule for each activity. An intermediate step is to construct what may be termed a logic network. This shows graphically, on a discipline basis, the inter-relationship of design tasks (including the groups of inter-dependent tasks) and information flows on a pseudo timeline (Figure 4). The construction of a bar chart programme requires durations to be assigned to tasks and also the number and duration of likely iterations around inter-dependent tasks. This raises issues that as yet have not been addressed. Hassan [8] argues that one approach for the production of such design programmes is the application of simulation techniques.

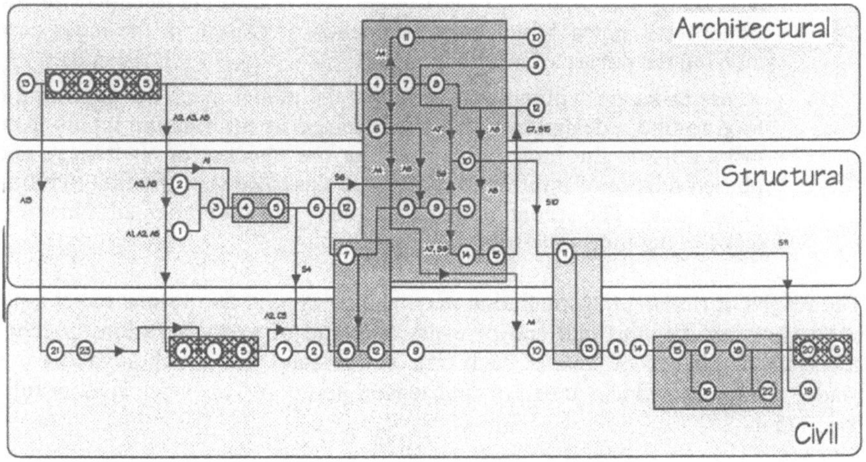

Figure 4 A logic network, Austin et al (1996)

4.4 Discrete Event Simulation

Hassan [8] uses data flow models and matrix analysis as a 'front end' to a discrete event simulation model, where durations and time factors related to the design tasks are then introduced. Hassan selected discrete event simulation as the most appropriate modelling environment after a review of potential techniques, then created a model with a proprietary simulation modelling product. This involved writing routines (in the program's own syntax) for all steps of the model building. The model was built up in modules that are created on the screen and combined to run the simulation.

The data required to build the simulation model are based upon the information links between the design processes of the data flow model, the grouping of design tasks within iterative loops having been identified from the matrix analysis. The main features of the simulation model are:

(i) it is able to run in either deterministic or stochastic mode. The user chooses from a pop-down menu, the stochastic mode involving sampling from a normal distribution;

(ii) links between different tasks may be switched 'on' or 'off'. A switch may be 'off' if a design task can proceed based on estimated data inputs. Subsequent tasks are tagged 'conditional'. When the design task receives the finalised information, a second iteration is carried out on the conditional design tasks, with a reduced duration based on a percentage of the first iteration;

(iii) it allows for phased release of information, which introduces an important refinement whereby a task with several outputs releases them in a pre-defined order, allowing some dependent tasks to start earlier;

(iv) it can simulate of 'gate keeping' of information, a communication problem associated with individuals who hold on to design data instead of making it available to the design team. Information links in the model have 'gates' which may be opened or closed, in the latter case delaying the release of information;

(v) where tasks are within iterative loops, the model maps the tasks in the loop to find a design task that has received its information inputs from tasks outside the loop (after carrying out checks for 'switches', and phased release of information), and can therefore start the first iteration of the loop. The second iteration of a loop can start when all remaining tasks in the loop fulfil all conditions.

The results of running the simulation may be displayed in two forms: either a bar chart showing the start and end of every task and of every iteration; or icons showing the change of state of each task as the simulation clock advances e.g. ready to start, started and at end of first/second iteration. This work is more fully reported in [17].

5. Potential Of The Techniques

Research has confirmed the suitability of DFDs for studying information transfer within the design process. Their production and use on live projects has helped the design team identify information which, although initially considered to be of low importance by some design staff, proved to be of great importance to others members of the design team. The source of critical data items became readily apparent and the nature of these items varied from project to project. Generic models of the conceptual/schematic and detailed design phases have identified and formally recorded the information requirements required to complete these stages of building design together with their relationship. Models of the detailed design phase are already being used by one major design & build organisation to provide checklists of information requirements and to assist in the general management of the design process. DFDs would therefore appear to have significant merit as method of defining the building design process and the information flows between design tasks; they are simple to create, easy to understand and do not impose an order on task execution (unlike critical path networks).

Analysis of the tasks within the blocks of a partitioned matrix shows a close correlation with many of the problems that blight the building design process. It is these multi-disciplinary task groupings that need careful management and co-ordination, and should be the units that are programmed (as opposed to single discipline operations). Studying the matrix can lead to a greater insight into the cross disciplinary co-ordination needs for particular sections of the building design process and reveals the most efficient order to perform design tasks. External factors, such as, resource availability, method of procurement or construction programmes may change this order and the effects can be identified by re-analysis of the matrix. Engineers have expressed the view that the design matrix not only allows them to make informed judgements but is also a potentially powerful visual tool to demonstrate to a client the effects of changes.

The value of a simulation model increases as the design progresses. No substantial benefit was found from running the simulation model to produce design schedules at the conceptual stage of design. At this stage of the design process architects 'think' on more than one task simultaneously; this aspect is impossible to simulate and the amount of time spent carrying out the design problem varies considerably. There is usually substantial waiting time for the client to make decisions, the resources used at the conceptual design stage are minimal and hence no substantial benefits accrue from simulating resource utilisation. The benefits become increasingly evident as the schematic design develops. It is later in the design process that benefits are achieved through assessing the quality of design information and the maximum benefits of this technique are envisaged during the detailed design process. The simulation tool has been found to be useful in that it gave to all concerned with the design process an understanding of the impact of the progress of work within other design disciplines. Senior design managers have stated that any models or techniques produced should not concentrate solely on the 'traditional' design disciplines. To fully achieve the benefits of new ways of working, input from other designers, specialist suppliers and client organisations needs to be considered.

Analysis and discussion on these techniques and their use in design organisations has highlighted the changing roles of the managers of design disciplines within the design process. How such 'change management' is controlled needs to be addressed. Early indications are that it may differ for each project depending on the type of work involved.

6. Conclusions

The construction industry, and building in particular, is currently focusing considerable attention on improving the management of the design process. The new IMI initiative to consider construction as a manufacturing process is leading both researchers and industry organisations to evaluate new methods, tools and techniques. Methods, tools and techniques that have been successful in the manufacturing industries are receiving particularly close attention.

Some researchers are currently focusing their attention on improving the management of the information flow within the design process. They consider that this approach will bring both immediate and lasting gains in efficiency and effectiveness. The development of new tools and techniques based upon DFDs' matrix methods and revised programming and scheduling techniques will precipitate these improvements and provide a platform for future research and development.

However, there is much other research work and experience in other industries which might have a bearing on the approach to the construction industry. Some of these approaches that have been identified are those of Hsu [20], Kusiak [21] and McCord & Eppinger [22]. In addition, Fisher [7] has pointed out that knowledge based systems may help reduce the differences between the construction industry and other sectors in terms of design, and hence assist the transfer of methods and philosophies. Closer co-ordination between researchers active in design management in other industries is therefore required if the most appropriate solutions for building design management are to be identified and implemented.

References

1. Newton A J 1996 The planning and management of detailed building design. *PhD thesis,* Loughborough University

2. Potter K, 1995 *Planning to design?*. (A practical introduction to the construction design process) a special publication report Number 113 published by The Construction Industry Research and Information Association CIRIA London, UK

3. Latham Sir M, 1994 *Constructing the team* . By HMSO London

4. Aqua Group 1990 *Tenders and Contracts For Building.* Blackwell Scientific Publishing Oxford UK

5. Chartered Institute of Building 1988 *Code of Estimating Practice Number Two: design and Build'* Chartered Institute of Building UK

6. Hainsworth D, 1996 The impact of design and build upon the construction industry. *a final year project report* Loughborough University UK

7. Fisher N, 1993 Construction as a manufacturing process? *Inaugural lecture* University of Reading

8. Hassan T M 1996 Simulating information flow to assist building design management. *PhD thesis,* Loughborough University

9. Topalian A, 1979 *The management of design projects* Associated Business Press, London

10. Bennet J, Flanagan R, Lansley P, Gray C, & Atkin B, 1998 *Building Britain 2001* a report from the Centre for Strategic Studies In Construction . Reading UK

11. Gray C, Hughes W, & Bennett J, 1994 *The successful management of design.* The University of Reading

12. Neale R, & Neale D, 1989 *Construction Planning* . Thomas Telford London UK

13. Barrie D S, & Paulson B C, 1992 *Professional Construction* Management. (Third Edition) Mc Graw Hill International New York

14. Austin S A, Baldwin A N, and Newton A.J, 1994 Manipulating the flow of design information to improve the programming of building design. *Construction Management and Economics* 12:445-455.

15. Austin S.A, Baldwin A.N, and Newton A J. 1996 A data flow model to plan and manage the building design process. Accepted for publication in *Journal of Engineering Design* 7:1: 3-25.

16. Austin S A, Baldwin A N, Thorpe A, and Hassan T, 1995 Simulating the construction design process by discrete event simulation. *Procs 10th Int. Conf on Engineering Design ICED95* Prague August pp 767-772

17. Baldwin A N, Austin S.A, Thorpe A , .and Hassan T, 1995 Simulating quality within the design process. *Procs 2nd American Society of Civil Engineers Congress for Computing in Civil Engineering* Atlanta June pp 1475-1482.

18. Steward D, 1981 Systems Analysis and Management. *Structure, Stategy, and Design* Petrocelli Books, New York

19. Eppinger S, Whitney D, Smith R, and Gebala D, 1990 Organising the tasks in complex design projects. *Proceedings ASME Conference on Design Theory and Methodology* New York pp 39-46

20. Hsu C, 1994 Manufacturing information systems in *Handbook of design manufacturing and automation..* Dorf R, and Kusiak A, (eds) J Wiley and Sons London

21. Kusiak A, 1994 Concurrent engineering; issues models and solution approaches in *Handbook of design manufacturing and automation* . Dorf R, and Kusiak A, (eds) J Wiley and Sons London.

22. McCord K, and Eppinger S,. 1993 Managing the integration problem in concurrent engineering. *MIT working paper*, Sloan School of Management WP 3594-93

Design as Building and Reusing Artifact Theories: Understanding and Supporting Growth of Design Knowledge

Jayachandra M. Reddy[1], Susan Finger[2], Suresh Konda[2], Eswaran Subrahmanian[2]

[1]Rockwell Science Center
Palo Alto, CA 94301 USA

[2]Engineering Design Research Center
Carnegie Mellon University
Pittsburgh, PA 15213 USA

Abstract

As artifacts are designed, knowledge is accumulated gradually and — as this knowledge is organized and reused — designs and design processes are continually refined. An understanding of the nature and growth of design knowledge and its reuse is essential for implementing better design systems and effective design practices. To develop such an understanding, we introduce *artifact theory* as an interdisciplinary theory about an artifact that is essential for designing that artifact. This theory encapsulates various types of synthetic, analytic and process knowledge and reconciles many disciplinary theories in the context of the artifact. We argue that it is necessarily a contextual theory and hence is ephemeral. While highly mature and well understood design domains may have complete artifact theories, in most domains artifact theories evolve during design. That is, designers not only produce a manufacturable description of the artifact, but also produce the corresponding artifact theory. We observe that this involves both adaptation and reuse of elements of existing artifact theories as well as development of new elements. Hence, we propose the view of design as building and reuse of artifact theories as the basis for understanding design and for developing design environments. We describe artifact theory in terms of several disparate views of design and bring them together leading to a unifying view. We discuss the implications of the view for computational design environments and outline our current research efforts in advancing and supporting this view.

1. Introduction

In designing an artifact, designers bring together knowledge from various sources for use in context as shown in Figure 1. The sources may include basic sciences such as physics and chemistry, engineering sciences such as thermodynamics and fluid mechanics, manufacturing and production sciences, empirical knowledge from handbooks and manuals, catalogs, previous designs, experience and creativity of designers, *etc.* A comprehensive design environment (computational or otherwise) should provide access to these various sources of knowledge, facilitate their use, help designers organize and reuse what they learn, and support sharing, collaboration and negotiation of design knowledge. The focus of this paper is on understanding the nature of design knowledge and its growth and the resulting implications for design environments.

The motivation for this work arises from a simple but important observation that engineering design not only involves knowledge use, but also knowledge building. This can be corroborated by the fact that even moderately complex design problems involve extensive argumentation and reconciliation, experimentation and understanding, development of new modeling assumptions and methods, and other activities typical of the task of knowledge building. While some elements of the knowledge required for design may remain invariant, often many elements are newly developed. However, most design environments are not evolutionary in that they do not support capture of newly generated knowledge and its reuse in future designs — at least not computationally. This paper presents a basis for understanding knowledge building during design processes and for developing design environments that can evolve with the accumulation of knowledge.

We introduce *artifact theory* as the encapsulation of knowledge about an artifact that forms the basis for understanding and supporting design knowledge and its growth. A theory is systematically organized knowledge applicable in a relatively wide variety of circumstances, especially a system of assumptions, accepted principles, and rules of procedure devised to analyze, predict, or otherwise explain

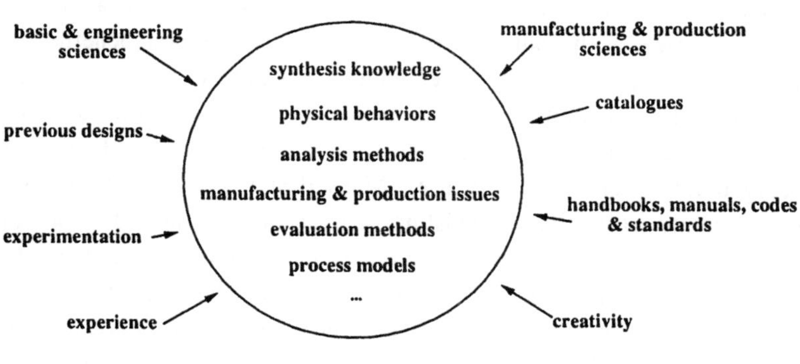

Figure 1: Designers use knowledge from many sources to design an artifact

the nature or behavior of a specified set of phenomena. In general, a theory should answer a variety of questions about the phenomenon under its purview. An artifact theory is a contextual theory that provides us with knowledge for designing and analyzing an artifact and for explaining and predicting the nature of the artifact. By contextual, we mean that the purview of an artifact theory is limited to a single artifact or a set of artifacts.

The outcome of a design process typically is viewed as a manufacturable description of the artifact consisting of detailed geometric models and drawings, specification of materials, list of parts and assembly specifications, *etc.* To reflect the knowledge building aspect of the design process, we extend this view and propose that design is a process of constructing a theory of the artifact, not merely constructing a manufacturable description (Figure 2).

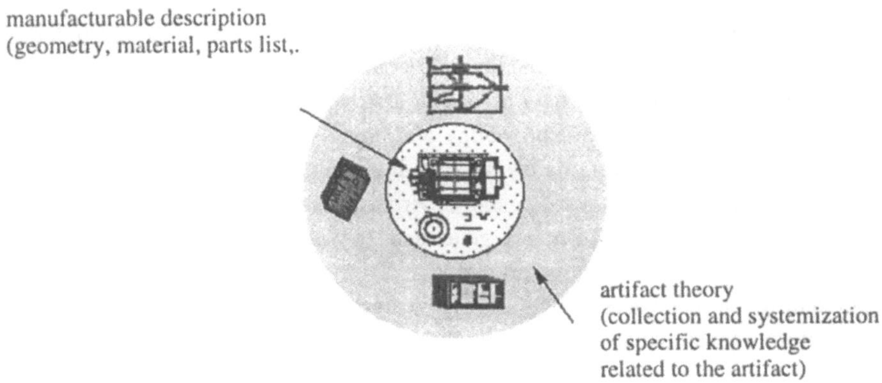

Figure 2: The outcome of design is an artifact theory which includes a manufacturable description

In general, artifact theory building involves both development of new elements of the theory as well as use and adaptation of elements of existing theories. This view of design as building and reuse of artifact theories and the shift towards expectation of artifact theories as the outcome of design provide several benefits to design practice. First, an artifact theory can explain the state of the artifact and can predict consequences of any changes. Second, artifact theories, or elements there of, can be reused or adapted for new designs. Despite this, there is a gap between support for theory building and for theory use. Many design systems take the view that design is mainly utilization of existing artifact theories and do not allow the building or evolution of theories. They ignore the activity of knowledge building, resulting in loss of the artifact theory that is so painstakingly developed during design. The view of design as building and reuse of artifact theories suggests that design environments should not only support use of existing theories, but should also support capture of new theories. With the addition of such theories, design environments should be constantly enriched and evolved.

The rest of the paper focusses on elaborating the need for contextual theories, advancing the view of design as building and reuse of artifact theories, and identifying its implications for design environments. We provide a philosophical and empirical basis for artifact theory and bring together several disparate views of design to establish a unifying theme. We then describe the nature of design process with respect to how knowledge is created, shared and reused in design, and identify its implications for computational environments. Finally, we raise some research questions and outline our current research efforts on elaborating and characterizing artifact theories, formalizing them in terms of languages, and supporting their development and use as a natural process of design.

2. Artifact Theory

What is a theory and what is its nature? We present a brief overview of scientific theories, especially with respect to their nature in design. There is abundant literature and debate on the definition, structure, and growth of scientific theories in the areas of philosophy of science and logic. Although some broad and general consensus exists, viewpoints differ. Readers interested in views presented by various schools of thought regarding this issue should refer to the critical introduction by Suppe [1].

2.1. Nature of Scientific Theories

Without delving into debates on the philosophy of science, we will discuss some essential functions and properties of scientific theories. Mehlberg summarizes empirical and theoretical aspects of scientific theories which we accept as the basis for discussion of theories [2]. He suggests that empirical aspects are determined, in substance, by a few essential functions discharged by any scientifically acceptable theory within the scientist's overall activity. He also argues that scientific theories could not discharge their essential functions without employing three formal components. These components provide a mechanism by which knowledge is put into storage and used at will, *i.e.*, they are the means by which a theory is formalized. The essential functions of a scientific theory are listed as below.

1. *Summarizing.* One of the principal functions of a theory is to summarize potentially large amount of information in a few condensed statements. For example, the theory of mechanics can be condensed into a single variational principle from which many mechanical laws can be derived.

2. *Predicting.* A theory should not only summarize established laws and facts, but also predict what may be established in the future. A theory must predict any fact which would be observed under any specifiable circumstances. For example, the theory of mechanics is expected to

predict mechanical laws which might be discovered.

3. *Explaining.* The essence of explanation consists in reducing a situation to elements with which we are so familiar that we accept them as matter of course, so our curiosity rests. A theory should also determine why a fact known to have taken place actually did so, why a law known to be valid in its proper realm of phenomena is actually valid there, *etc.*

4. *Controlling.* A theory should enable us to bring about desirable changes in our environment by following procedures which the theory indicates.

5. *Informing.* A scientific theory should provide us with relevant and dependable information about objects which are observable. The dependability of information is due to its being supported by the outcome of other investigations. In other words, scientific theories should provide us with empirical knowledge.

The following are the three formalisms by which a scientific theory discharges the essential functions listed above.

1. *Logical Formalism.* Using the logical formalism, the elements of science are built one upon another to generate new knowledge — the reverse of the process of explanation [3]. The axioms and definitions of a theory are manipulated using logical formalisms to generate theorems of the axiomatic system. Substitution of different logical formalisms in a theory may give different predictions or explanations.

2. *Mathematical Formalism.* Mathematical formalisms are essential to discharge empirical functions. Examples of mathematical formalisms are Hilbert spaces (in quantum mechanics) and linear algebra (in structural mechanics). The difference between logical formalism and mathematical formalism is that substitution of different mathematical formalisms in a theory does not affect the explanations or predictions of the theory.

3. *Metaphysical Formalism.* Scientific theories are based on some assumptions or axioms that are undecidable on logico-mathematical grounds and are not susceptible to any observational test. This class of axioms forms the metaphysical formalism of a theory. For example, in dynamic theory of gases, one of the axioms is that gases consist of rigid molecules that fly about in all directions, colliding with each other and with the containing wall.

While these empirical and theoretical aspects provide a basis for evaluating a theory, all generally accepted scientific theories do not discharge all the functions equally well and do not have all three theoretical formalisms. Typically, a theory will discharge some functions better than others.

2.2. Reconciling Theory and Practice in Design

Philosophy of science strives to understand the nature and structure of scientific theories, including their roles in the growth of scientific knowledge. The subject of philosophy of engineering, which is not as well established as philosophy of science, addresses similar issues in the context of engineering. It addresses questions such as what is the nature of theory in engineering design practice and what is engineering knowledge? Addis presents the views of various philosophers of engineering and weaves the views together to establish the nature and role of theory in design [3]. He addresses the question of whether the normally accepted classification of theory and practice is relevant in the context of design, specifically in structural engineering. Most engineers agree that a gap exists between theory and practice; however, Addis argues that there are several possibilities of gaps, as shown in Figure 3, and that it is often unclear which gap is under discussion.

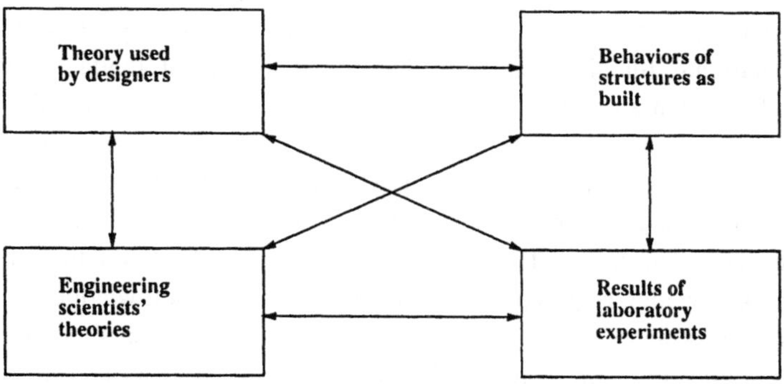

Figure 3: Gaps between theory and practice in design [3]

An engineering science is closer to a pure science such as physics, and its ultimate aim is to understand and explain the phenomena under its purview. Both engineering and pure sciences make use of theories for such explanation. On the other hand, engineering design is concerned with the production of artifacts in conditions less predictable and less under control. The aim of engineering design and engineering science are quite different. Engineering design uses what is typically called engineering knowledge, which may consist of empirical data, rules, laws, intuition, design procedures, experience and codes of practice. While engineering science and theories are useful, for example in establishing empirical rules and in analyzing the artifact, they are rarely directly useful in designing. Engineering knowledge does not fall easily under either category of theory or practice.

If we wish to bridge the gap between theory and practice, *i.e.*, make practice the application of theory, we must extend our definition of theory to include not just engineering science but all engineering knowledge. Addis argues that this extended definition of engineering theory discharges its functions in ways similar to scientific theory. The role of a theory in physical sciences is to explain certain phenomenon, whereas its role in design is to aid in the transformation of functional specifications to a solution that can be manufactured. Although the role of engineering knowledge in design is well recognized both in design research and practice (exemplified by the applications of knowledge-based systems), there have been relatively few efforts to formalize this knowledge in terms of theories.

2.3. Artifact Theory as a Contextual Theory

With this extended definition of theory for design, which includes engineering knowledge, let us ask the question: Is there a theory that is useful for the purpose of designing any given artifact independent of the context? Before answering this question, let us briefly examine the nature of artifact theories in terms of its empirical aspects. These include informational, explanatory, predictive, constructive (or controlling) and summarizing functions which are listed in Table 1. What elements should an artifact theory consist of to discharge these functions? The informational function demands that the theory must contain and deliver relevant information about the artifact. This information may include the structure of the artifact, its geometry, functions, and so on. The explanatory and predictive functions are similar and demand that the dependencies between information be captured in the theory (for example, what requirements influenced a particular design decision and how). The constructive function demands that the theory capture the knowledge required to design the artifact. This knowledge may include knowledge about analysis methods, synthesis methods, and evaluation methods. The summarizing function demands that the theory be compact and make explicit knowledge that can be reused across several different artifacts.

A general theory for design should encompass all the engineering knowledge, concisely or otherwise, and should be able to deliver its functions with respect to all artifacts. Ideally, a general theory should be able to transform given functional specifications into realizable physical attributes independent of the context, the same way that general theories in physical sciences (*e.g.*, general theory of relativity) explain much of the universe. However, no such well trenched theories exist for guiding the design process. Later in this paper, we describe some efforts in developing theories for design and analyze them with respect to the empirical and theoretical aspects of scientific theories.

In the absence of general theories for design, a pragmatic approach is to consider developing and using more contextual and specific theories for designing specific artifacts. An artifact theory is such a contextual theory that is essential to design and build a specific artifact or a class of artifacts. This notion of artifact theory can be exemplified by the existence of theories such as the theory of bridge construction and horology (the science of making time pieces). If we look at the history of the growth of theories in physical sciences, we notice that more unifying and more

Function	Elements	Examples (from motor design)
Informational	Information about the artifact: structure, functions, geometry, ...	• What is the composition of the cooling system? • What are its functional requirements?
Explanatory & predictive	Explanation for the state of the artifact Prediction of effects of any changes	• What requirements caused the selection of symmetric cooling flow? • What happens if requirements become more stringent?
Constructive	Knowledge about design methods, analysis methods, *etc.*	• How do I go about designing a motor with certain specs? • How do I simulate motor thermal characteristics? • How do I evaluate the performance of the motor? How do I benchmark it?
Summarizing	Patterns in the artifact information model	• What are possible configurations for a cooling system? • How do I adapt the motor thermal model?

Table 1: Elements of artifact theory and their functions

general theories came only after several disparate unconnected theories which were first developed to explain different physical phenomena. Therefore, starting with the development of contextual theories is a reasonable approach toward developing a more general understanding of design.

Contextual theories are developed by acquiring and systematizing knowledge in context. This may lead to the formation of layers of artifact theories reflecting various levels of contextualization. The theories at lower levels are formed by borrowing and adapting theories at higher layers and adding additional context dependent knowledge. Figure 4 shows the role of context in the organization of theories. While the most fundamental sciences, such as physics and chemistry, are constituents of all artifact theories, more contextual theories exist for specialized classes of artifacts, more specific theories, *e.g.*, the theory of design and construction of electric motors, are formed by borrowing and adapting general theories and adding additional context-dependent knowledge. Even more specialized theories may exist for a class of electric motors, like AC induction motors. The most specific theories are about individual instances of artifacts. Different artifact theories may overlap and contain several common elements that can be transferred and reused across artifacts.

3. Design Studies

To understand building and reuse of artifact theories, we have undertaken three participatory case studies. The first concerned the design of a third generation AC induction motor for hybrid electric vehicles. This case study was an experiment in distributed design between Carnegie Mellon University and Stanford University [4].

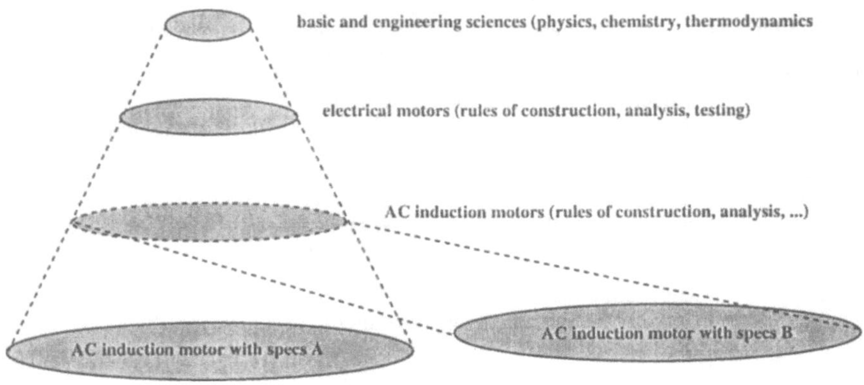

Figure 4: Hierarchy of artifact theories

Since two previous generations had been designed and manufactured, we could study the evolution of artifact theories over generations. The second case study concerned the design of a laser tracker which was part of the Madefast experiment [5]. The third project was the wearable computer design project at the Engineering Design Research Center at Carnegie Mellon [6, 7].

In the design projects studied, designers used many categories of knowledge including personal knowledge, knowledge from previous designs, knowledge about physical behaviors, and manufacturing knowledge. A detailed analysis of knowledge use in the motor design project is presented in [4]. One of the key observations regarding the design process relates to the large amount of information generated. For example, the documentation generated during the design of the AC induction motor is several hundred pages long. The final report describing design specifications, solutions, experiments, simulations, decisions and rationales is about 300 pages long. Since this design had two years of history, the previous documentation was used extensively. A significant portion of the design effort involved accessing and interpreting previous design information. The sought-after information was not only about final design drawings, but also about simulation models, underlying assumptions, experiments and their results, and manufacturing processes and constraints. Embedded within the information in design reports is knowledge that designers seek in designing other similar artifacts. A good design report is a compilation of the theory about the artifact (albeit non-computable); use of these reports exemplifies reuse of elements of existing artifact theories.

In these design case studies, designers did not have enough knowledge *a priori* to transform functional specifications into physical attributes. In practice, it is hard to implement the idealism of function to form mapping via behavior. A significant amount of contextual knowledge must almost always be invented. While this is clearly true for novel designs, the requirements for knowledge are unstable even in mature domains. This instability arises from the variety of requirements imposed by context, such as new customer requirements, new technologies, *etc.* For example, in a case study of the design of transformers in a large company, Finger *et al.* show that although design of transformers is over 100 years old, there is still significant novelty in the designs [8]. In related work, McMahon identifies several modes of incremental design including parameter space exploration, improved understanding of explicit-implicit attribute relationships, change in product design specification, modification of feasible design space, and change in the design principle [9]. Researchers working in the area of learning and design reuse have also looked at how knowledge from prior designs must be rationalized and made explicit [10, 11].

Designers must seek out knowledge in the context of the design problem, debate the relevance of this knowledge, and generate new knowledge. As suggested by Vincenti [12], such knowledge development may involve introducing blind variations, experimenting, and selectively keeping what works. Vincenti also states that variations are more likely to be tried out vicariously by analysis and experiment in place of direct trial in design of mature products. Another significant mechanism for the growth of engineering knowledge is the occurrence of design failures, when designer's conjectures are falsified [13]. Both Vicenti's and Petroski's models of knowledge growth were observed in the case studies described above.

Designers are aware of the instability of the design knowledge needed; Meyer, who developed a formal grammar for architectural and structural design of tall buildings, presents critiques of his prototype grammar by two domain experts [14]. The first expert criticizes the grammar because it will never be able to capture and formalize the knowledge behind design of tall buildings completely since the knowledge is always changing. In other words, artifact theories keep evolving.

> "It's a hopeless task. The minute that you've got it you'll want to do something else... The problem of designing a building is more like writing a poem than writing a sentence. The subtleties of that are so immense that you can't hope to duplicate it... The structural engineering of buildings, I think, is like being a builder of armature for sculpture. The stuff that's inside the Statue of Liberty is Eiffel's. To try to write a program that would anticipate every sculpture that a sculptor could come up with would be sort of a big waste of time, I think, although somebody might argue." [14]

However, the second expert realizes that it is desirable to formalize what we already know about the design process.

> "Other than these comments, I think you are on the right track. I like the idea of developing a grammar to establish the entire vocabulary, to establish a set of rules of how you can operate within the process." [14]

4. Disparate Views & Reconciliation

Several disparate views of design have been proposed in the past to guide design practice as well as development of computational environments. In this section, we briefly describe some of these views and identify common themes relating to artifact theory. The purpose of this section is to elaborate on artifact theory from the perspective of different views and bring them together, leading to a unifying theme of design as building and reuse of artifact theories.

General Design Theory (GDT). GDT [15] is a theory of design based on axiomatic set theory. It states that, given ideal knowledge, design is a mapping process from function space to attribute space with no substantial computation required. This theory, being too idealistic to model real-world design process, has been extended with the adaptation of ideal knowledge to real knowledge [16, 17]. Design with real knowledge requires the ability to continually model the designed artifact until it evolves into a set of candidates that satisfy the given specification. However, GDT, both in the ideal and real knowledge, only applies to domains with known topological structure. Thus, Reich hypothesizes that GDT may be most applicable in established design domains with well-developed categorizations and accumulated knowledge about how to mediate between the function and the attribute categorizations [18]. Designing in these domains may range from simple selection from a catalogue, to composing systems from available components, to a stepwise refinement process.

Our notion of artifact theory is similar to ideal and real knowledge in GDT. Ideal knowledge includes knowledge about topologies of functional spaces and attribute spaces as well as mappings between them. The real knowledge includes knowledge about how to mediate between functional and attribute spaces, *i.e.*, both to propose refinements as well as to evaluate them. GDT tells us that these are the elements of knowledge (*i.e.*, artifact theory) required for design. GDT assumes the existence of topological spaces prior to design, knowledge about mappings between them, and knowledge about how to mediate between functional and attribute spaces. In reality, complete knowledge rarely exists *a priori*, and often a significant part of the knowledge is invented during designing of an artifact. In terms of GDT, our view of design suggests the task of designers is to put together the ideal and real knowledge required for the development of the artifact. Through repeated design of similar artifacts, this knowledge can crystallize into ideal knowledge.

Design as a Computable Function. Fitzhorn proposes that the design process can be modeled using the abstract model of computability: the Turing machine [19]. In his model, designs are enumerated strings from a possibly multidimensional grammar, and design specifications or constraints are formal state changes that govern string enumeration. He defines a Turing machine in which the state transition functions are dynamic. This adaptation does not to alter any characteristics of general Turing machines. The design process is then mapped onto the dynamic-state Turing machine. The design process is a series of state transitions from initial specifications to final artifact description via transition functions, which are constraints on the design. The transition functions themselves are dynamic. The design includes three distinct, but interrelated elements: the design process, the artifacts of design, and a set of design constraints. The process is independent of the design context. The artifacts and constraints are domain dependent, and he assumes they can be formally defined.

In this model of design, the formal models of artifacts (strings) as well as transition functions on these models are context specific; these context-specific models and transition functions form the artifact theory. To use this model of design as the basis for implementing design systems, one must assume the availability of artifact theories, *i.e.*, formal models of artifacts and transition functions that transform these models. These assumptions typically do not hold true in reality, and development of these formal models and transition functions itself is a central part of the design process. As Fitzhorn points out, the exposition of design as a computable function tells us more about what cannot be done than what can be. However, it formalizes the knowledge required for designing facilitating better understanding of artifact theories.

Theory of Technical Systems. Hubka and Eder present a comprehensive and unifying theory to promote the understanding of technical systems [20]. They extend the definition of theory in the context of design in a manner similar to that discussed in Section 2.2. The term *technical system* means all human-made artifacts, including technical artifacts and processes. The primary aim of the theory is to classify and categorize knowledge about technical systems into an ordered set of statements about their nature, regularities of conformation, origination, development, and empirical observations. The content of the theory includes many aspects of technical systems: their constituents, structures, and the models that describe them; properties of technical systems and their inter-relationships; evaluation of technical systems from the viewpoints of engineers, users, and society; formal representation of technical systems (such as drawings); and historic evolution of technical systems. Hubka and Eder argue that the lack of a general meta-theory results in the unsystematic collection of know-how. The theory of technical systems is intended to be such a theory which brings together independent domains and generalizes them. Hubka and Eder distinguish between general theories and special theories of technical systems: many special theories provide concrete statements that are derived from the general theory. As shown in Figure 5, the specialized theories are enriched by cross-currents and cross fertilization from various types of machinery and from established scientific theories such as thermodynamics and fluid dynamics.

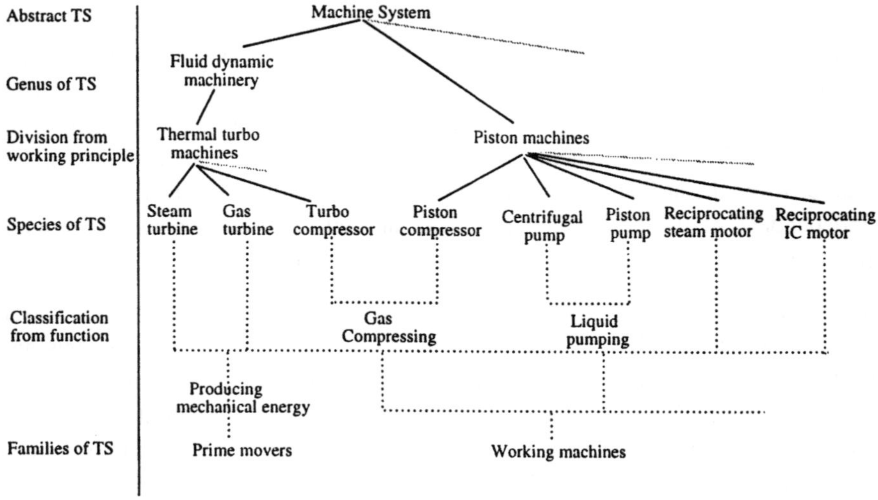

Figure 5: Formation of types of technical systems from [20]

In a way, a special theory of a technical system is equivalent to an artifact theory, both in its insistence on context and in its role. Hubka and Eder describe and characterize several elements of a general theory of technical systems, concrete elements of which are developed within the realms of specialized theories, in other words, artifact theories. However, they do not share our view of design as theory building; they propose the theory of technical systems as a means to collect and organize knowledge about engineering artifacts for use.

Shared Memory in Design. Konda *et al.* present a unifying theme for design theory emphasizing the importance of context [21]. They argue that a theory of the design process should be constructed that emphasizes the empirical and descriptive aspects of design, moving away from universal, context-less prescription. They present the concept of shared memory as the embodiment of both context and shared meaning and propose that the emphasis should be placed on developing shared memory. They distinguish between vertical shared memory and horizontal shared memory: vertical memory encapsulates increasingly detailed aspects of a given profession's knowledge, whereas horizontal memory encapsulates the record of interdisciplinary communication. Horizontal shared memory always requires mutual translation of terms and concepts across groups, because members of design groups working on the same design do not share the same experiences, concepts, perspectives, exemplars, methods, or techniques. Creation of shared memory requires a deeper understanding of how individuals create shared meanings, what special languages, representations, models, *etc.*, are required to enable different individuals to share knowledge and experience. The authors also argue that design systems should not be limited to aiding an activity, but should also focus on creating communication channels to facilitate development of shared memory.

Artifact theory can be seen as a kind of shared memory because it embodies context and shared meaning among designers. Artifact theory is an encapsulation of knowledge about an artifact (or a class of artifacts), *i.e.*, it is tied to the artifact whereas the context of shared memory is unspecified — artifact theory, in a way, makes shared memory concrete.

Artifactual Engineering. Recently, Yoshikawa has proposed a new field of study called *artifactual engineering* [22]. The purpose of creating a new field is to bridge the gap between specialized engineering domains (*e.g.*, mechanical and electrical) that have evolved separately. Yoshikawa argues that we need to establish an engineering discipline that denies the existence of domains; he calls it *artifactual engineering*. He argues that as long as engineering is portioned into domains and depends on traditional territorialized principles, the goal of artifactual engineering will be to show us facets that are not normally visible.

> If we try to build up process theories in artifactual engineering now, we ought to learn from the history whereby structural dynamics was established by referring to knowledge needed in making bridges and building before structural dynamics came into being. The lesson is that it is necessary not only to produce a massive amount of products but also to consider industry, which has produced vast amounts of knowledge, an important source of wisdom. This knowledge has not been systematized, and formally it is the same type as the knowledge required for making a delicious omelet. If this can be called primitive knowledge, then in artifactual engineering it may be possible to attain research methods aiming to establish process theories, including general information about the degree to which systematic knowledge using it is organized into effective knowledge via abduction. [22]

Artifactual engineering shares concepts with theory of technical systems and shared memory in design. Although artifactual engineering denies the existence of specific domains (as currently established in engineering), its relation to the role of context is unclear. However, similar to our view of design, artifactual engineering emphasizes the study of artifacts and their design processes as opposed to specific engineering domains alone. In our view, generalization of artifact theories to form more general theories is the essence of artifactual engineering.

Axiomatic Design. Suh proposes two axioms for design, namely the functional independence axiom and the information minimization axiom [23]. These axioms can be considered to form the metaphysical basis for a theory of design. While this theory has a metaphysical formalism (*i.e.*, the axioms of design), it lacks logical and mathematical formalisms, *i.e.*, how to design from such axioms logically and algorithmically. While the axioms themselves are useful, further work must be done in developing logical and mathematical formalisms to complement them. These formalisms can be developed within the context of specific artifacts leading to development of contextual artifact theories based on Suh's axioms of design.

These contextual theories should interpret the axioms in the context of specific artifacts and provide design procedures based on them.

Programming as Theory Building. In the domain of software engineering, Naur argues that computer programming is equivalent to constructing a theory about the program [24]. Naur argues that programming is primarily building knowledge of a certain kind as an auxiliary product. This knowledge is generated, in part, through matching some significant part of an activity in the real world to the formal symbol manipulation done by a program running on a computer. Naur describes how programmer's knowledge is encapsulated into a theory. Having such a theory enables programmers to explain, justify, and answer queries about their program. This notion of "theory about the program" is similar to our notion of artifact theory.

5. Building, Sharing & Reuse of Artifact Theories

To support design as artifact theory building and use, we need to understand the nature of the design process in terms of how knowledge is created, shared, and used. In this section, we argue that the artifact theory is necessarily shared among the design participants and is developed collaboratively by the team. This has an important implication that design systems can support capturing of elements of theories by supporting and capturing systematic discourse. Based on this, we propose a model of design process that can be the basis for design environments.

5.1. Artifact Theory in Design Discourse

We assert that the artifact theory is shared among participants in the design discourse. While there may be some tacit elements in design knowledge, we believe that much of this knowledge can be expressed, captured, and communicated. Designers working on the same design must be able to express the theory of the artifact to one another. While design knowledge generated by a single designer may remain implicit, what needs to be communicated will be expressed and made explicit. This assertion can be verified by the content and quantity of information shared during the design discourse in the case studies described earlier.

This assertion can be verified by the content and quantity of information shared during the design discourse in the case studies described earlier. The theories developed and shared in discourse typically reside in various communication tools and are expressed in many forms. We define design discourse as the interactions between participants in the design — both human designers and computer tools. We assert that the artifact theory is shared among participants in the design discourse. Among these participants are clients, designers, manufacturers and suppliers. Human designers deal with ill-conditioned issues of the problem that are understood well enough to be incorporated in computer tools. Computer tools deal with issues that are well understood and can be formulated in specific computational methodologies.

Several researchers have explored various aspects of design discourse and placed discourse at the center of design process. Bucciarelli, based on ethnographic studies involving an X-ray system design, describes the style of design discourse and presents three illustrations: constraining discourse, naming discourse and decision discourse [25]. The first is about setting of performance specifications early in the design. The second is about naming, which is a design phenomenon that crystallizes images of parts and functions of the design in the minds of participants. The third is decision making. All three, *i.e.*, setting requirements, naming, and decision making, can be seen as the outcome of an argument. An example of a model of discourse is IBIS [26], in which designers propose, criticize, refine, abstract, and make concrete ideas and concepts that lead to a final product. Konda *et al.* argue that design is an activity in which designers move toward a shared understanding of the design artifact by negotiation and reconciliation of several different perspectives [21]. The view taken in this document is that designers construct a theory of the artifact collaboratively through negotiation and reconciliation of different perspectives and interests. From the case studies, the following properties of design discourse can be observed.

- *Information Content.* Large quantities of information are generated and communicated in the discourse among the design participants. In [4], we present gross quantitative measures of information generated and communicated during the AC induction motor design. We observe that the discourse changes as design moves from defining and finalizing specifications to detailing a manufacturable description of the artifact.
- *Tool Usage.* For reasons arising from the need for efficient information exchange and from the diversity of participants, design information is transmitted in a variety of representational forms [27]. Different media are suitable for different needs and representational forms, and hence different media are used for modeling and communication of design information.
- *Knowledge Intensiveness.* Design discourse is knowledge intensive, *i.e.*, much knowledge is exchanged in discourse. In previous studies, we have presented analyses of patterns in knowledge use in the evolution of design in the case studies and show that much of it is in deed embedded in the discourse [4, 28].

5.2. Model of the Design Process

In this section, we present a model of the design process that incorporates theory construction, theory sharing through discourse, and use of existing theories. This model, shown in Figure 6, is a basis for developing computational environments. The model consists of several participants including designers and computer tools, the artifact theory being constructed, a collection of existing artifact theories that are available for reuse, conceptual knowledge, and external knowledge. Designers' characteristics include competency in generating and evaluating design solutions, ability to acquire and assimilate conceptual knowledge, ability to communicate with other designers, and ability to use and build elements of artifact theories. Computer tools may be used in the design process to perform specific tasks, for example, a finite element analysis tool for thermal analysis. While these tools may be independent of the artifact, their use in the artifact design (usage models) is included in the artifact theory. These usage models may include analytical assumptions, assumptions about appropriate representations for the artifact, and so on. For example, when a general mathematical tool such as MATLAB is used to perform simulation, the modeling methods and assumptions underlying the dynamic model are a part of the artifact theory. Conceptual knowledge refers to the knowledge associated with individual designers, which is not codified and hence not accessible to others. External knowledge refers to any knowledge accessed and used by designers outside the design environment.

The process of theory construction involves several tasks, including reconciliation of different perspectives, conceptualization and generation of ideas, reuse of existing theories, and experimentation. This necessitates several interactions between the elements of the model shown in Figure 6. Reconciliation of perspectives requires that designers communicate with one another. This communication may require sharing the theory itself, represented by the interactions between the designers and the artifact theory. The interactions between the designers and the existing artifact theories reflects access and reuse of these theories.

Other interactions are shown with dashed arrows which may include interactions through synchronous communication such as face-to-face meetings, telephone and

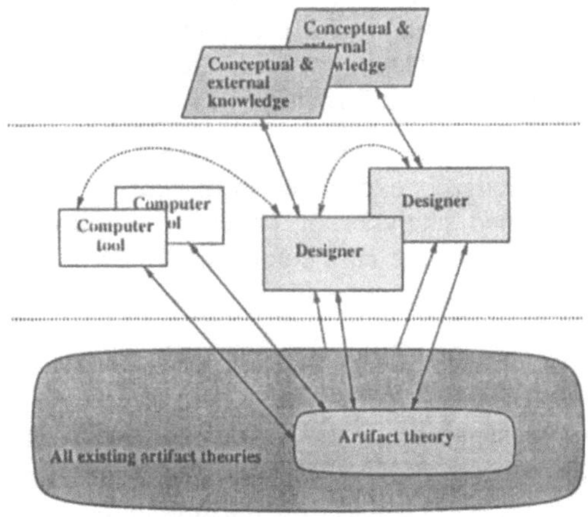

Figure 6: Model of design process

video conferences, application sharing via computers, and shared whiteboards. Asynchronous communication may be by e-mail and faxes. Other interactions included interactions between designers and tools and between designers and conceptual and external knowledge. Conceptualization of knowledge requires understanding of the context — both the context about the artifact and what is available in the world that is relevant. This requires that relevant information be available on as-needed basis. Recent advances in information technologies can benefit enormously in this regard — information on manufacturing processes, catalog parts, and so on can be made readily available.

6. Implications for Design Environments

Although design involves artifact theory building, the outcome of design is usually not seen as the theory, but rather only as a manufacturable description. Of the four elements of a theory listed in Table 1, only the first and part of the second are usually considered to be the outcome of design and most current design environments conform to this view. However, the foremost implication of the view of design as building and reuse of artifact theories is that design environments should support explication of artifact theories in computational forms as well as their use in future designs. In effect, they should bring together and integrate the theory building and theory use aspects of design. Whether or not support is provided for explicit theory building, theory building is inherent in the design process. Since these theories necessarily are shared among design participants, by supporting design discourse in a structured fashion, design environments can assure explicit artifact theories as the outcome of design.

In general, computational tools used in a design process can be categorized into two broad categories: those that primarily support use of existing knowledge and those that primarily support creation and sharing of new knowledge through discourse. The former category of tools have certain knowledge about the artifact and support a specific design activity. For example, in the motor design studied, a tool that performs thermal analysis of an AC induction motor has embedded in it knowledge about AC induction motors. The latter facilitate the design process, communication, argumentation, *etc.*. These support articulation and sharing of newly discovered knowledge, for example, word processors, e-mail, and documentation tools such as PENS [29] which provides limited support for structured modeling.

A gap exists between those tools that support theory use and those that support creation and sharing of theories (Figure 7). This is because the theories shared in the latter category of tools are not immediately visible and are not immediately incorporated into the former. A closer relationship between the two categories of tools will make the design processes more efficient and productive. Ideally, design environments should facilitate articulation of the knowledge developed during design in computable forms that is explicit and that can be readily used.

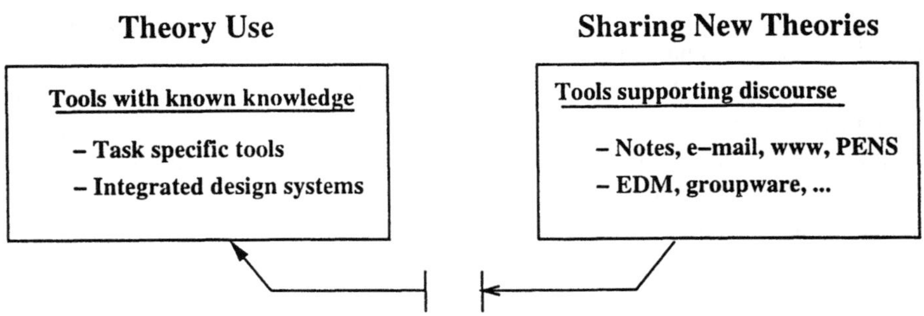

Figure 7: Gap between theory building and theory use

The view of design as building and reuse of artifact theories has some important implications for development of integrated design systems. Integrated design systems result from well developed and consolidated artifact theories. To integrate task-specific tools in the context of designing of an artifact, we need to know the interactions between the tasks, for example, the mappings between vocabularies of different tools. For example, in the AC induction motor design project, the tools used include AutoCAD for drafting and geometric design, MathCAD for thermal analysis model solving (resistance model) as well as for fluid dynamic analysis, Algor for finite element thermal analysis of end turns, and machining planning software tools at the job shops. The questions in integrating these tools include: do we have knowledge about interactions between these tools? If we know the interactions and incorporate them in an integrated design system, how stable are our assumptions about these interactions? The answers depend on the dynamic characteristics of the artifact design — how often the technologies change, how much and how often the context and requirements change, *etc.* The less dynamic the artifact design is, the more likely our assumptions about interactions remain valid, making integrated design systems possible. Unfortunately, many unpredictable interactions often occur between designers and disciplines, and therefore concurrent design systems cannot connect together discipline-specific tools in predefined ways [30].

Distributed agent architectures for concurrent engineering systems have been proposed, for example [31], where the focus is on integrating existing systems via shared models of knowledge. The claimed advantage of the agent-based model of collaboration is that computational resources can be utilized opportunistically and added incrementally [32]. The fundamental problem of information sharing and development of shared vocabularies (ontologies) still remains a bottleneck. It usually takes considerable effort to build such ontologies to make an agent a player in problem solving. Such ontologies have been developed and demonstrated in specific applications — however, the generality of such development processes and feasibility of the overall philosophy of plug and play is still questionable. Olsen *et al.* [32], based on experiences with agent-based architectures, conclude that information technology must address a basic need — the need for collaborators to establish information sharing agreements and to incorporate these agreements into the tools they use. These observations leave us with several questions: when is integration of tools possible and when not? It is conceivable that we can integrate tools at the level of sharing detailed geometric information, for example using PDES/STEP standards — but to what levels can we raise this integration? What aspects of design can be integrated and what cannot be?

7. Research Questions

In order to both practice and support design as building and reuse of artifact theories, several questions must be answered which include: What is the content of an artifact theory? Can we formally represent and capture it, and if so how? How can we transfer elements of artifact theories across artifacts? How can we support theory building and use in computational environments? In this section, we outline our current research efforts in answering these questions. A more complete description is given in [28].

What do designers look for in documents and reports concerning previous designs and what do they transfer across designs? In addition to looking for answers to specific questions, we believe they look for patterns in artifact information and transfer them across generations of artifacts and across different artifact types. These patterns are major constituents of artifact theories. The artifact information consists of information about composition of the artifact, *i.e.*, how various concepts or subsystems compose the artifact. For example, the motor designed is composed of a stator, a rotor, a specific type of cooling system, *etc.* Artifact information also consists of descriptor models that describe various aspects of the artifact and its subsystems or subconcepts. For example, the motor has a thermal model, an electromagnetic model, and a design process model, among others. The patterns both in artifact composition and in descriptor models of the artifact capture significant knowledge about the artifact. The patterns in artifact composition capture synthesis knowledge, *e.g.*, knowledge about what concepts can be combined well and how they can be combined. The patterns in descriptor models capture knowledge about specific aspects of the artifact. If these patterns can be made explicit during design, they become visible and can be reused in similar designs. Our assertion is that that design environments should support not only modeling of artifact information but also explication and reuse of patterns in artifact information.

To develop computational environments that support theory construction and use in terms of patterns, we need appropriate computational representations for patterns in the artifact information as well as for artifact information itself. We propose that the artifact information can be represented in terms of generalized graphs, and patterns can be represented in terms of grammars on these graphs. This is true for both patterns in concept composition as well as descriptor models. Posing the theory in terms of formal grammars is appropriate since grammars can capture knowledge effectively as corroborated by many knowledge-based applications in engineering design, for example [14]. Grammars also facilitate formal analysis and characterization of artifact theories and development of metrics for design such as complexity metrics and similarity metrics.

A computational environment can cohesively support theory building and use by providing support for explicating patterns in artifact information in the form of grammars and for using these patterns effectively in future design and modeling activities. In [28], we study theory construction and use in terms of grammars using an environment based on *n*-dim [33]. This environment supports concurrent development of models of artifact information as well as grammars underlying them. It also provides several features required for collaborative modeling including information management, access control, negotiation, *etc.* thereby facilitating the discourse through which artifact theory is shared and consolidated.

8. Conclusions

Universal theories for design do not exist; that is, we do not have theories that are all-encompassing and that provide support in designing any artifact of choice. In the absence of such universal theories, a pragmatic approach is to consider developing and using contextual theories for designing specific artifacts. We have introduced *artifact theory* as such a contextual theory. We have argued that design involves development of not only a manufacturable description of the artifact but also its theory. This theory building often involves both creation of new elements as well as use of elements of existing theories. This view of design — as building and reuse of artifact theories — provides a framework for understanding the nature and growth of design knowledge. This view, as a guiding principle for design practice, can benefit organizations significantly in building their respective corporate memories. However, in correspondence with this view, design environments should support structured discourse and structured modeling of design information making the underlying knowledge explicit leading to computable and reusable artifact theories.

Acknowledgements

This work is supported by the Engineering Design Research Center at Carnegie Mellon University, an Engineering Research Center of the U.S. National Science Foundation, under Grant No. EEC-8943164.

References

1. Suppe F 1977 *The Structure of Scientific Theories.* University of Illinois Press, Urbana IL

2. Mehlberg H 1962 The theoretical and emperical aspects of science. In: Nagel E, Supes P, Tarskied A (eds) *Proceedings of the 1960 Congress on Logic Methodology and Philosophy of Science.* Stanford University Press, Palo Alto CA, pp 275-284

3. Addis W 1990 *Structural Engineering: The Nature of Theory and Design.* Ellis Horwood Limited, West Sussex England

4. Reddy J M, Chan B, Finger S 1996 Patterns in design discourse: a case study. In: *Knowledge Intensive CAD, Volume 1.* Chapman & Hall, London, pp 265-283

5. Cutkosky M, Tenenbaum J, Glucksman J 1996 Madefast: an exercise in collaborative engineering over the internet. *Communications of the ACM* 39(9):78-87

6. Siewiorek D P, Smailagic A, Lee J CY, Adl-Tabatabai A R 1994 Interdisciplinary concurrent design methodology as applied to the navigator wearable computer system. *Journal of Computer and Software Engineering* 2(3): 259-292

7. Finger S, Stivoric J, Amon C. et al. 1996 Reflections on a concurrent design methodology: A case study in wearable computer design. *Computer-Aided Design* 28(5):393-404

8. Finger S, Gardner E, Subrahmanian E 1993 Design support systems for concurrent engineering: A case study in large power transformer design. In: *Proceedings of The International Conference on Engineering Design ICED'93.* The Hauge, pp 1433-1440

9. McMahon C 1994 Observations on modes of incremental change in design. *Journal of Engineering Design* 5(3):195-209

10. Duffy A H B, Kerr S M 1993 Customised perspectives of past designs from automated group rationalisations. *Artificial Intelligence in Engineering* 8:183-200

11. Duffy A H B, Duffy S M 1996 Learning for design reuse. *Artificial Intelligence for Engineering Design Analysis and Manufacturing* 10:139-142

12. Vincenti W 1990 *What Engineers Know and How They Know It.* John Hopkins University Press, Baltimore MD

13. Petroski H 1985 *To Engineer is Human: The Role of Failure in Successful Design.* St Martin's Press, New York

14. Meyer S A 1995 Description of the structural design of tall buildings through the grammar paradigm. PhD thesis, Carnegie Mellon University

15. Yoshikawa H 1981 General design theory and a CAD system. In: *Man-Machine Communication in CAD/CAM.* North Holland, Amsterdam, pp 35-58

16. Tomiyama T, Yoshikawa H 1986 *Extended General Design Theory.* Tech Report CS-R8604 Centrum voor Wiskunde en Informatica, Amsterdam

17. Tomiyama T, Yoshikawa H 1985 Extended general design theory. In *Design Theory in Computer-Aided Design* North Holland, Amsterdam, pp 95-130

18. Reich Y A 1995 Critical review of general design theory. *Research in Engineering Design* 7(1):1-18

19. Fitzhorn P A 1994 Engineering design is a computable function. *Artificial Intelligence in Engineering Design and Manufacturing* 8(1):35-44

20. Hubka V, Eder W E 1988 *Theory of Technical Systems: A Total Concept Theory for Engineering Design.* Springer-Verlag, New York

21. Monarch I A, Konda S L, Levy S N, Reich Y, Subrahmanian E, Ulrich C 1993 Shared memory in design: theory and practice. In: *Social Science Research Technical Systems and Cooperative Work.* Paris, pp 227-241

22. Yoshikawa H 1992 Proposal for artifactual engineering: aims to make science and technology self-conclusive. In: *Illume A Tepco* Semiannual Review Tokyo Electric Power Co Inc., Tokyo, pp 41-56

23. Suh N P 1988 *The Principles of Design.* Oxford University Press, Oxford

24. Naur P 1985 Programming as theory building. *Microprocessing and Microprogramming* 15:253-261

25. Bucciarelli L L 1988 An ethnographic perspective on engineering design. *Design Studies* 9(3):159-168

26. Rittel H, Webber M 1973 Dilemmas in a general theory of planning. *Policy Sciences* 4:155-169

27. Hubka V (ed) 1991 *Proceedings of ICED'91 International Conference on Engineering Design.* WDK, Zurich

28. Reddy J 1996 Building and reuse of artifact theories: A view of design and its implications for computational environments. PhD thesis, Carnegie Mellon University

29. Hong J ,Toye G, Leifer L 1995 PENS: personal electronic notebook with sharing. In: *Fourth IEEE Workshop on Enabling Technologies.* Berkeley Springs West Virginia

30. Finger S, Konda S, Subrahmanian E 1995 Concurrent design happens at the interfaces. *Artificial Intelligence for Engineering Design Analysis and Manufacturing* 9:89-99

31. Cutkosky M R, Engelmore R S, Fikes R E et al. 1993 PACT, an experiment in integrating concurrent engineering systems. *Computer* 26(1):28-37

32. Olsen G, Cutkosky M, Tenenbaum J M, Gruber T 1994 Collaborative engineering based on knowledge sharing agreements. *Proceedings of the ASME Database Symposium.* Minneapolis MN pp 11-14

33. Levy S, Subrahmanian E, Konda S, Coyne R, Westerberg A, Reich Y 1993 *An Overview of n-dim Environment.* EDRC Technical Report 05-65-93 Carnegie Mellon University

EDD'96 Programme Committee

Chair:
Dr A H B Duffy (UK)

Vice-chairs:
Prof. M M Andreasen (DK)
Prof. S Finger (USA)
Dr L T M Blessing (UK)

International Advisory Board:
Prof. H Birkhofer (D)
Prof. D C Brown (USA)
Dr C Hales (USA)
Prof. L Leifer (USA)
Prof. K J MacCallum (UK)
Prof. N F M Roozenburg (NL)
Prof. T Smithers (E)
Prof. T Tomiyama (J)
Prof. S Vajna (D)